My THINKING CAP

Solutions for Global Crisis

Brad Fregger

Austin, Texas

My Thinking Cap
By Brad Fregger
© 2008 Brad Fregger

Groundbreaking Press
8305 Arboles Circle
Austin, TX 78737
(512) 657-8780
www.groundbreaking.com

Library of Congress Control Number:
ISBN: 0-9793542-6-9

First Edition

Editor
Barbara Foley

Book Design & Production
M. Kevin Ford

Cover Design & Production
Original Front Cover Image
M. Kevin Ford

All rights reserved. No part of this book may be reproduced or utilized in any form by any means, electronic or mechanical, including photocopying or recording, or by any information storage and retrieval system, without permission in writing from the author.

Dedication

To my beautiful and talented wife Barbara Foley, to whom I owe so much. This book could not have been written without your unwavering belief in me and support of my ideas. I love you!

Acknowledgements

This book is the result of years of learning how to think and how to effectively consider solutions being offered for societal issues, both big and small. I have had the advantage of wonderful teachers throughout my life, teachers who have shown me the wisdom of looking carefully at all alternatives and then considering carefully what alternatives might have been missed.

It goes without saying that those closest to me were my first and most important teachers. I was blessed to be born into a family of relatively healthy individuals, both physically and mentally, so I didn't have to overcome the negative influence found in severely dysfunctional relationships. My first marriage blessed my wife and I with three wonderful children (and now their equally wonderful spouses, seven terrific grandchildren, and one great-grandchild), who in many ways have been my greatest teachers.

I now have a beautiful and talented wife who is an editor *par excellence*. To tell the truth, this book is as much her work as it is mine. Somebody was watching out for me.

And then, of course, there were my mentors—life has blessed me with quite a few. I won't try to list you all for fear that someone important might be omitted. Thank you one and all. A piece of each of you can be found in this book.

Preface

I have been asked, "Who do you think *you* are, suggesting to the world's leaders what needs to be done to assure the success of our civilization?" I must admit that, at times, I have wondered that myself. ...

The truth is, I don't see much wisdom coming from our "leaders." In fact, I often see a total lack of wisdom in regard to some critical issues. I have met many people who have a great amount of authority and responsibility, and I have met many thousands more who are doing the real work of the world. ... I don't see a difference when it comes to the knowledge needed to get the job done, or the wisdom needed to know what needs to be done. I am very confident that we have some exceptional individuals serving us in Congress and throughout the government, but as a group ... let's just say ... they leave a lot to be desired.

Much of the problem lies with our system of government, which is not a representative democracy, but a democracy run by special interests, at the whim of those interests. Add to this the bias of the media toward bad news (discussed in Chapter II) which has a direct, negative impact on the socio-economic health of our society, probably our civilization. The fact they take no responsibility for this, only exacerbates the situation.

By the way, it's my total lack of faith in our government to make the right decisions at the right time, and its inability to get the job done, that makes me laugh at conspiracy theories. For example, hypothesizing the involvement of our government in 9/11 ... no way that could be pulled off ... anyone who believes it could be, is woefully ignorant of the organization that has to be in place to make something like that happen, and the inability of any part of government to carry it off. The truth is, believe it or not, it is much easier for a small group of committed individuals to carry off a 9/11 attack than it would be for any government in the world. In the U.S. especially (due to our freedom of the press) it would be impossible for anything of that magnitude to be kept a secret.

Bottom line is, I decided that I had as much right to share my solutions as anyone does—and I mean anyone. In addition, because of my academic learning and professional training, I just might have something of value to offer; and, if I do, I not only have a right, I have a responsibility to share my thoughts and solutions concerning global crisis issues. Something I say may give to someone who does have authority the idea or inspiration they need to create the right solution.

I got my Master's in Futuristics (societal futures) in 1980, but I've been considering these issues since long before then. I actually decided to get my Master's because I wanted to add significant academic work to the personal research that I had been doing for over twenty years.

This book contains some essays that were originally written almost thirty years ago. Surprisingly, very little has changed. However, I have updated all of the essays to be sure that they are relevant to what is going on in the world today.

Each chapter stands on its own, even though they sometimes relate back to what was said in a previous chapter. The first two chapters describe important philosophical issues that will help you understand the basis for many of my beliefs. These "introductory" chapters are followed by four of my most current essays, all of which discuss issues that are currently under consideration by both the public and government agencies. I have attempted to title these chapters very clearly, in order that the Table of Contents will be of some help.

Also, my style of writing is conversational. I dislike any style of writing that makes it more difficult than necessary for the average person to understand what it is you're trying to convey. Ultimately the responsibility for communication lies with the communicator ... if my reader is having a problem understanding me … I must look to myself first.

That's all for now. I *trust* that you will judge the time you spend with this book to be time well spent. It is my *hope* that you will be galvanized to share with others those solutions that seem reasonable to you. After all, that's one of the main reasons that I wrote this book.

Secondly, it is also my hope that I have inspired you to think more deeply about all or some of these issues, and to consider what actions you might take to be a part of finding solutions to these areas of global concern.

Brad Fregger
Austin, Texas
July, 2008

Foreword

Brad Fregger is, as they say, "a piece of work"! I'm not sure where that expression came from but I mean it as a compliment, and I think it is usually intended as one. What I'm trying to say is: my husband is a go-getter, a can-do character, and one happening guy! And on top of that, as I am fond of telling him, he's ridiculously happy most of the time. Once in awhile he indulges in an hour of depression or allows an odd wave of paranoia to come over him, and it feels literally to me like the sun has gone behind the clouds. I am always so relieved when it passes and my ray of sunshine returns.

    I asked Brad's mother what he was like as a child and she said, "Very high energy! If he had been born today, they probably would have put him on one of those drugs, but of course, I was a young, inexperienced mother and didn't quite know how to handle him—and I never could handle him. Luckily he never got into any serious trouble."

    She went on to tell me that he ventured on his first "trip around the block" when he was three years old. His dad managed a movie theater in Longmont, Colorado at the time; the theater was several blocks from their home. One afternoon his mom got a call from Rolly, his dad, "Martha, do you know where your son is?"

"Well, yes, he's out playing in the yard."

"Are you sure? Go see if you can find him." So Martha went to look outside.

When she found no trace of little Brad, she went back to the phone, "I can't find him. He's nowhere in the yard."

Rolly responded, "Well, I didn't think you would find him because I happen to be looking at him right now! He's out here on the sidewalk, in front of the theater, with his little red wagon."

So Martha figured if he could find his way to the theater, he could easily find the corner grocery, which was only a couple of blocks away. She started sending him off with his little red wagon, a few dollars, and a list, to pick up whatever groceries she needed. He was ready, willing, and able, and it never occurred to her that he might come to harm. Luckily he never did, and just as in his adult career, he always "delivered the goods" in a timely manner.

I could tell you more stories about Brad's childhood, but the point is that he has always been eager to expand his horizons. He likes to think about the big picture and come up with solutions to people's problems. I guess it all started with solving his mom's problems by venturing out to the local grocery with his little red wagon. And he's been pondering, and venturing, and solving ever since!

All that ambition and natural curiosity to explore his surroundings and beyond has given Brad quite an interesting career path. He went from managing menswear stores; to corporate training and development; to being the first producer of computer games and inventor of computer solitaire; to having his own software development company; to helping with several other startup ventures; to being a college professor and public speaker, as well as running his own book-publishing company.

I could tell you more, but you won't believe me! Several years ago, the wife of one of the senior officers of a startup that Brad was newly involved in, invited him over to their home for dinner. She encouraged Brad to talk about himself and, always eager to have an audience, Brad related many of his past career adventures and accomplishments. She wanted to hear more, so Brad willingly shared one after another of his varied career twists and turns with her and her husband. It got late and finally the evening ended.

About a year or so later, when she had gotten to know Brad better, she told him after that first evening, she was convinced that he was full of hot air. She didn't think there was any way one person could have possibly done all those things. Of course, by that time, she knew that everything he had shared with them that night was true.

I am amused, and yet amazed, when Brad tells me he feels like his life truly began when he "came home" to Austin and me in August 1996. While I feel that is a very romantic thing to say, it strikes me as a bit ridiculous because of everything he had already experienced and accomplished up to that point, which was quite a lot! But somehow life in Austin has made it easier for him to appreciate his success on the professional level and to genuinely savor the accolades, which come his way. If I've been able to contribute to that in any way, I am deeply grateful, because I do believe Brad is a brilliant, talented individual with much to share and a contagious zeal for sharing. It's truly a privilege to have him as a life partner and be able to witness his joy and success, as well as his pure passion for living. He is a constant inspiration to me.

When we married in December 1998, he promised me two things: that he would always love me, and that life would never be dull. So far he has certainly kept those promises. We started our married life with the understanding that I would make the little decisions, like where we would live, what we would eat, what clothes we would buy, where we would go on vacation—you get the idea; and he would make the big decisions, like who should be president, how to solve the oil crisis, how to deal with the border problem, what to do about climate change—again, you get the idea.

So, you see, Brad has been busy making "the big decisions" and that is what this book is all about. Being a writer, he has been putting his ideas about "the big decisions" down on paper for many years now, and it has surprised him to discover that essays he wrote even thirty years ago, are still as pertinent as they were the day he wrote them.

I agree with Brad that these ideas and proposals are worth sharing and I have encouraged him to publish them in this book you are holding. Hopefully his words and ideas will act as a catalyst to get you thinking about some or all of these issues as well. Too many of us Americans are comfortable and spoiled. We take so much about our lives for granted, and we assume that we will always enjoy the quality of life we have today. Hopefully that will be the case, but we have no right to make such assumptions.

It's time to think what we can do, right now, to make sure that our American way of life survives for our children, grandchildren, and great-grandchildren to enjoy. It's time also to think about what it means to be a citizen of the world and to learn to think globally.

It's time for all of us to put on our thinking caps! But, first, let's give Brad our undivided attention.

Barbara E. Foley
Austin, Texas
June 17, 2008

Table of Contents

Acknowledgements v
Preface vii
Foreword xi
Table of Contents xvii
1 - The Image of the Future 1
2 - The Power of the Story 9
3 - A Solution for the Energy Crisis 29
4 - What's New about Climate Change? 57
5 - A Solution for the Immigration Crisis 73
6 - What is Intelligent Design? 89
7 - Effective Leadership 103
8 - The Enlightened Company 131
9 - Achieving Innovative Environments 145
10 - Streamlining the Budgeting Process 163
11 - Defeating Elitism 181
12 - Business Ethics in Crisis 187
13 - Embracing the Unexpected 203
14 - Our Future in Space 219
Author's Bio 231

I
The Image of the Future

This chapter, based on the subject of my Master's Thesis, "The Impact of Images of the Future on Current Socio-Economic Health," was originally written in the late 70s and updated in 1998—nothing of significance has been added since then, nor needed to be.

In the past few years it has become a widely accepted belief that the world we live in is running out of the essential resources needed to support our civilization. Crude oil and all forms of energy are in limited supply. There is great concern over both the deterioration of the ozone layer and global warming (or cooling, depending on the current "sky is falling" fad). While the threat of war between super powers isn't as likely as it once was, there continues to be the threat of limited conflicts that could escalate into a confrontation involving other nations, including our own. And there is always the threat of a relatively small group of committed individuals using terrorist tactics to disrupt and possibly destroy the foundations of our society. In addition, growing food to feed all people now alive, as well as millions more to be born, for some, does not seem to be a realizable goal.

There is also the tension between those who are concerned for the survival of humanity and our way of life, and those who are concerned for the survival of the ecology as a whole; too often the goals of these two groups seem to be diametrically opposed. Because of these and other factors, a wave of pessimism has been sweeping the world for the last couple of decades.

Influential authors continue to write of potential doom for our way of life. This call has been carried by scholars, scientists, leaders in government and industry, science-fiction writers, and many private environmental organizations and concerned citizens (not to mention the media which thrives on news of impending disaster).

A central theme in many of these arguments is: the world cannot continue on its present course because to do so can only lead to the breakdown of our economic system and ultimately the death of millions of people. Most of the alternatives offered do not contain within them ways to save our civilization, but rather suggest ways that small groups of us might be able to weather the coming storm.

If we are to believe these authors, there seems little we can do to escape a future that holds no hope for untold millions of people destined to live lives filled with physical and emotional suffering, and ultimately death, as the wheels of our civilization grind to a halt.

Is there any hope? Can we, as a society escape this pessimistic future and instead begin the Twenty-First Century with hope and faith as we enter a new age of plenty, where the challenge of preserving our world and its ecology becomes a goal consistent with the survival of humanity and our way of life?

There are many authors who see much to be hopeful about, authors who believe that technology holds a hope for the future and that we may discover ways to overcome what seem to be impossible odds from the vantage point of our currently limited knowledge. Others believe that we are on the brink of experiencing a new level of consciousness that will lead to the greatest surge of knowledge and potential the world has ever seen.

What does the future hold for our society and for other societies on the Earth? Will the actual future be closer to the doom we all fear, or to the promise we hope for? What kind of control do we have over the future? Is there anything we can do?

I believe there is. Our hope for the future lies in a better understanding of how our vision or image of the future impacts the future itself, and how this knowledge can be used to create for our society, a positive future with hope and potential for generations of humanity.

The Image of the Future

The future of our society is dependent on our hopes, expectations, and images of it. I do not mean the long-term future, years or even decades ahead, but tomorrow, the short-term future only days, weeks or, at the most, months from now. We can increase the possibility of a more positive future by developing and nurturing positive visions and images of it.

F. L. Polak in his book, *Images of the Future*, said, "Human society and culture are being magnetically pulled towards a future fulfillment of their own idealistic images of the future, as well as being pushed from behind by their own realistic past." Our thoughts and beliefs of what the future will be like can move us toward that future and help to create and affect the present.

While a strong, positive vision of the future does lead to confidence, I'm not talking only about a confidence in the future. Confidence is a reaction to a belief about the coming short-term future, next month or next year. No, what I'm talking about is a people's response to a vision of what the future can be. An image of a future that is as far reaching and challenging as any of the great visions of the past, the type of vision that inspires a society and the people within it. When the people begin to believe a vision of this magnitude, everything is changed and the world is never the same again.

O. W. Markley of the Stanford Research Institute said, "The development of our societies has been in large part dependent on the creative vision of the great members of the human race, and of the images of the future of the important cultures throughout history." Malachi Martin, a Jesuit priest and student of why societies succeed or fail, in his book *The New Castle* said, "The greatest heights of culture and civilization were always fashioned, not within the molds of sociological cause and effect, but within a transcendent vision." Finally, the late Ed Lindaman, a past director of program planning for the design and manufacture of the Apollo Spacecraft and past president of Whitworth College in Spokane, Washington said, "The believer not only interprets history but, above all, he changes it, because he believes, because he hopes."

Malachi Martin, too, believes in the power of the image of the future. In his book he speaks of the image of the future as a vision of the "Castle." Different people and different societies throughout history have experienced this vision of the future, of a better world, a world where anything and everything was possible. The Castle was their vision of that ideal state.

For example, there was an obscure Semite nomad, named Abraham, who was the first witness of the Jewish vision and the patriarch of all of the major religions of the

The Image of the Future

western world. The power of his vision has lasted for over 5000 years.

How many of you have seen the desert around Salt Lake City, Utah? I don't know how you feel, but I have never seen one more desolate. The legends say there were only seven trees in that whole valley when Brigham Young chose it as the site for the Mormon community. They had no money and very little material goods. Most of them didn't even have horses. But they did have a vision, a vision of a city by the shores of the Great Salt Lake. Only seven years after the first person stepped foot in that area, the foundation to the temple was being laid; and today we have Salt Lake City, Utah, a thriving, green, and well-planned city, a jewel in the desert.

The vision, or image of the future is a society's conception or belief in a time that is yet to come. It can be either positive or negative, but, in either case, it must be an exaggeration of what would be reasonably expected. It is either a hope or a fear that lives in the minds of the societal group, both collectively and individually. When the image of the future is projected clearly, it is leading the society into a future that is truly their conception of tomorrow. It is a promise of what the future will be like within the lifetime of the society. It is my hope of what that future can be like within my lifetime. In other words, the society expects to see the actualization of the vision; the individuals within the society believe they will play a part in the realization of the vision, and they, as individuals, hope to live to see it happen.

There have been many visions that have shaped the history and future of our world including the "American dream" that is just slightly over 200 years old. Each of these cultures was/is being driven by a vision, a vision that began with a few people and spread throughout the society, shaping the future and transforming the present.

I feared the future, feared my children's world,
the outcome of the massive changes taking
place in our country.

My imagination runs wild with visions of the
breakdown of society,
millions dead and dying,
cities empty and decaying,
our way of life gone for centuries.

The future like a steamroller
rushes madly downhill,
no one at the controls.

Doubt and fear of things to come
still haunt me ...

As I move forward into the unknown,
Hope beckons.

The Image of the Future

I now believe that we have the knowledge and skills necessary to effectively handle every global crisis ahead of us. But, we must choose to use the resources that we have available, the natural resources, as well as the mental and spiritual resources. The questions are: Will we make reasonable decisions in regard to all of our resources? And, will we make the right choices in time?

This opening chapter discusses the role of images of the future to the socio-economic health of a society, and the importance of positive, challenging, visionary images. The next chapter discusses how these images are shared generation to generation—the power of the story.

2
The Power of the Story

The power of the story to change the way people perceive an issue, whether it be at the personal or societal level, is well known. What isn't well understood, is the mechanism that enables this to take place. This chapter contains excerpts from my book, *One Shovel Full: Telling Stories to Change Beliefs, Attitudes, and Perceptions*.

In the first chapter we talked about the importance of images of the future to the socio-economic health of a society, and that these images can be either positive or negative. In the following chapters I will, at times, point an accusing finger at the media, because their fixation on "bad news" is creating a negative image of the future—in large part, the global crises we are experiencing are due to the stories that are being told.

I'm not saying that they are doing this on purpose. I don't believe that the vast majority of the media has any sense of the impact their pervasive negative assault is having on our nation's, and the world's, socio-economic health. Likewise, I believe that the world's media has little understanding of how the mechanism they are using unconsciously creates the belief in many that western civilization, especially the United States, is responsible for most of

The Power of the Story

the world's problems. The alternative, that they have purposely set out to destroy western civilization, is unthinkable.

The mechanism I am talking about is the use of stories to develop and/or change attitudes, perceptions, and beliefs.

The human race is a species that loves stories, stories of all kinds. Since the beginning of human time, stories have been used to create the beliefs that are the foundation of its societies. Elders have seen it as their responsibility to educate the youth on what is important; what is worth fighting for—even worth dying for. This has been, and still is, critical to our survival as a social animal.

In the past, most of the stories handed down to the next generation have been stories of courage and passion held forth against all odds; stories of survival, success, and persistence. These stories made you proud to be a part of something bigger than yourself.

Certain segments of society, from the military to the gangs, from the church to the corporation, from the school to the home, still use stories to tell us who we are and what we believe. The result: we are proud to belong and we are committed to the group and its reasons for existence.

However, today the stories that are being told at the societal level are often not ones that foster a sense of pride and ownership. In fact, they tend to be exactly the opposite, fostering a sense of shame and unworthiness. These stories come from all facets of the media, from news and entertainment television, to the movies, to the songs we listen to and sing along with.

In this chapter I will present my theories of how and why stories are so powerful, and how they develop and change our ways of thinking. Once you understand the how-and-why of the power of stories, you will be even more concerned about how we are currently choosing to use them.

Much of the time, an individual, team, or society is not performing up to expectations because of a belief, attitude, or perception that is an obstacle to its achieving success. As the founder of three corporate training and development departments, my major objective was to determine how to teach individuals new skills as quickly and efficiently as possible, and to change their attitudes about a variety of issues.

This chapter's conclusions come from my long-time fascination with the mind and how it operates, and also from many years of observing how people learn.

Many in the training profession believe that attitude problems result from a lack of skills and/or knowledge. They say poor attitudes exist because people don't

The Power of the Story

clearly understand, at the expected level, what their job is or how to perform it—or both.

My experience has been that more often, especially with competent individuals, the root cause of behavior issues is attitude problems. While this may not be the case when people are placed in new jobs without receiving the support they need, my experience proves that changing attitude will change behavior.

The years I spent considering this issue have resulted in a hypothesis as to how the mind works. The hypothesis will seem somewhat simplistic to those experts who have looked more deeply into the scientific principles behind the mind's operation; however, like any good hypothesis, it explains things very well.

The subconscious controls what we perceive.

The foundation of my hypothesis is that the mind is made up of two major areas, the conscious and the subconscious. The conscious mind is responsible for everything we are aware of; the subconscious handles everything else and is completely responsible for what we perceive. This is the critical part: the subconscious controls what we perceive.

The five senses don't filter out anything. They don't have the capability; they are only capable of receiving data and passing it on to your brain. Due to the sheer volume, it is literally impossible for the conscious mind to handle all of the sensory data that the subconscious is continually receiving.

The subconscious mind receives the data supplied by the senses, analyzes it to determine how critical it is, filters out that which is not critical, and brings critical information to the attention of the conscious mind.

How does the subconscious decide what is critical and what isn't, what to filter out and what to allow to surface? I suspect there are many different reasons the subconscious decides to filter or not filter. And, I'm sure some of the reasons relate to the current focus of the conscious mind and others to the survival of the individual.

The subconscious mind has a strong tendency not to allow us to perceive those things that would greatly threaten our current belief system.

I suspect that this conscious/subconscious relationship is somewhat codependent, with the subconscious being very careful not to confuse the conscious mind. In this sense, it works very hard to protect the conscious from conflicting points of view. I am convinced this is a critical survival characteristic.

The Power of the Story

Most of us would agree that to succeed we must believe, and often failure results from doubt. For this reason, the subconscious mind has a powerful tendency not to allow us to see that which would threaten our strong beliefs.

The problem is, very often, it is critical that we see things from an entirely new perspective; that we have the ability to change our current perception of reality. Well, we do. ... It's called insight, or inspiration. We wake up one morning and ask ourselves, "Why have I been doing that?" Or, "Why have I been believing that?" And ... we change.

This new realization, insight, or inspiration can come from many different sources, but at this time, we are only interested in one of the sources—our subconscious mind.

The subconscious mind "mines" the data it receives for unknown relationships and then delivers that information, at the appropriate time, to the conscious mind as realization, insight, or intuition.

In January of 1999, I became VP of product development for Austin's Dryken Technologies, a data mining company with research offices in Knoxville, Tennessee. I was picked because I'm good at creating product, not because I'm expert in the area of data mining. In fact, when I took the job, I didn't have a clue as to what data mining was all about.

My first task was to get to Knoxville as soon as possible and have a long learning session with the data-mining scientists. The chief scientist, Dr. Nancy Grady, was still working at Oak Ridge National Labs at the time.

We sat down across from each other at the conference table and I said, "Nancy, you're going to have to start at the beginning. I don't know anything about data mining."

"No problem," she replied. "Basically it works like human intuition. Our mind takes in a tremendous amount of data continually; data about everything we experience, in every way we experience it. It categorizes the data and determines relationships of which we are not consciously aware. Then at the appropriate time, it feeds only specific, relevant information to us in the form of intuition. Data mining works the same way. The algorithms look at massive amounts of disparate data, determining relationships that could never be determined by human analysis; and then they report out those specific relationships."

I sat there in silence for a minute. If I understood what she was saying, this answered a multitude of questions I had entertained over the years. Finally, I looked over at her and said, "Let me tell you a story about myself; you tell me if it fits what you just described."

She nodded for me to go ahead.

"As the manager of a men's store in San Jose, California, I developed a very interesting talent. I would watch a customer enter the store and walk toward me. I was

The Power of the Story

usually standing in an area of the store about one hundred feet from the front door. When the customer reached me, I'd say, 'The Shoe Department is down that aisle to your right,' or 'Can I help you find a suit?' or 'Looking for a gift?'

"The customer would often look at me and say, 'How'd you know what I was looking for?'

"I couldn't answer ... I didn't know how I knew; I just seemed to know. I wasn't right all the time, but I was right often enough to make me wonder what was going on. Part of me wondered if an angel was standing beside me giving me this information, but I had trouble with this explanation. Angels must have more important things to do than tell me that John Smith is looking for the Shoe Department.

"I think you've finally given me the answer I've been looking for. From what you've said, it seems to me that from the moment John entered the store, my subconscious was categorizing and analyzing everything he did, where he looked, what he reached out and touched, how quickly he walked, what he was wearing, and on and on. It had done this hundreds of times as customer after customer entered the store and then made a request. At some point, through this natural data mining process, my subconscious figured out that people who behaved as this one was behaving usually were looking for the shoe department, or the suits, or for a gift ... whatever.

"Then I would receive an intuitive thought, a thought stimulated by my subconscious but determined by the data my senses had provided as the customer walked through the store to where I was standing. Right?"

"Right!" Nancy responded. "That's exactly what happened and exactly what data mining is all about. Our algorithms look at massive amounts of disparate information and discover, through the use of neural networks and other advanced data mining technologies, relationships that cannot be determined in any other way."

This was fascinating … I had sat down to learn about data mining and discovered invaluable information about the human mind and how it operates, what intuition is and how it works.

So how does all of this relate to storytelling?

Even though the subconscious mind is in complete control of what we perceive, it gets its cues regarding what information to deliver and what to filter out from the conscious mind. It's almost as if we were saying to the subconscious, "Don't show me that stuff … I don't want to see it!" The subconscious "hears" this command and then obeys, filtering out the information we would find disturbing.

In the same way, we let the subconscious know what information we believe is valuable, and therefore should be kept for future use, and which information is worthless and should be tossed. The end result: We see what we want to see and are "blind" to the rest!

The Power of the Story

The problem is: how to get critical data into the subconscious without letting the conscious mind contaminate it, especially data that runs counter to what we strongly believe.

When we can accomplish this, the subconscious can then use the data as part of its analytical process and at the appropriate time, provide us with a realization, insight, or intuitive thought, potentially changing beliefs, attitudes, and perceptions.

One of the first times I became aware of this potential was in 1980 when, as the director of training and development for the Atari Corporation, I looked at a system for changing attitudes called Neuro Linguistic Programming (NLP). Richard Bandler and John Grinder developed NLP as a result of intense research into exactly how effective counselors changed attitudes and beliefs.

Once they had determined the differences between successful and unsuccessful counselors' behaviors, they taught those different behaviors to the ineffective counselors and, as if by magic, they became significantly more effective. Their foundational book on NLP is titled *The Structure of Magic* because they couldn't explain how it worked, only that it did.

One of the major distinctions between the two groups of counselors, the difference that most interested me, was the way effective counselors used stories to make a point. It wasn't that they used stories, but *how* they used stories.

The effective counselors' storytelling process spoke directly to the subconscious, without the conscious mind being aware of what was going on. In other words, they were able to get data that ran counter to the conscious mind's current belief system—critical data—to the subconscious without it being contaminated by the conscious mind. Then the subconscious categorized and analyzed this new data, delivering the resulting new information to the conscious mind as breakthrough insight. This caused a change in the individual's attitude toward the specific situation.

Surprisingly, effective counselors did not explain their stories. In fact, in most cases, they went out of their way not to share why they were telling the client a particular story. When ineffective counselors told a story, they spent almost as much time explaining why they were telling it and what they wanted the client to learn from it, as they did in telling the story itself.

If we understand how the conscious and subconscious minds work together, we see why the effective counselors' stories had an impact and the ineffective counselors stories didn't. The ineffective counselors were begging, in fact, demanding the conscious mind to get involved. They were operating under the mistaken belief that the individual had to make a conscious decision to change. The truth seems to be exactly the opposite: an individual only changes his or her attitudes and beliefs through a subconscious process.

NLP isn't the only source that supports this supposition. For example, we know Jesus seldom, if ever, explained the parables. I wondered if He, too, understood this basic principle.

I emailed my son Jon, a Presbyterian minister, asking, "Are you aware of any times that Jesus explained a parable? If He didn't explain the parables, do you know why He didn't?"

Jon responded, "Yes ... Jesus did explain one parable. It is found in Luke 8:4-15. I think it is interesting that in verse 10 Jesus talks about why he uses parables. At the end of verse 10 He said,

> To you it has been given to know the secrets of the kingdom of God; but to others I speak in parables, so that "looking they may not perceive, and listening they may not understand."

"Jesus is quoting from Isaiah 6:9-10.

> And he said, "Go and say to this people: 'Keep listening, but do not comprehend; keep looking, but do not understand.' 10 Make the mind of this people dull, and stop their ears, and shut their

eyes, so that they may not look with their eyes, and listen with their ears, and comprehend with their minds, and turn and be healed."

"By quoting this passage Jesus is not just using the words; He is interested in using their meaning also. In Isaiah the people are made to be dumb on purpose. This is done so that the normal ways of gaining comprehension—eyes, ears, mind—cannot be used. A different kind, or level, of understanding is needed. This "other" learning is needed so that the people may '... turn and be healed.'

"What it is saying is that to experience real transformation ... the kind that turns your life around and brings about real healing ... it is best *not* to rely on our normal senses and ways of learning. It *doesn't* happen when we use our conscious minds ... but when we use the subconscious."

So ... stories have very real power to change what people believe, their attitudes, and their perceptions. And, they can be very effective, on both a personal and societal level.

Leaders and opinion makers have used stories forever to change how people think and what they believe. Some recent, specific examples are: the weapons of mass destruction in Iraq; the global-warming caused rise in sea levels that will kill and/or displace millions of people;

The Power of the Story

the feared extinction of the polar bear—how it can't possibly survive when global warming melts all the ice; the seemingly all-pervasive danger of stranger abductions; etc.

There are a million more examples of how stories impact each of us, what we believe, and what might happen if we don't quickly work to change the situation. When we realize that the story we were told was a fabrication, then we get very angry and we will probably never trust that person, or group again.

The best example of this is the public's perception concerning the weapons of mass destruction ... once they were not discovered, the majority of U.S. citizens turned against Bush and there's little chance that they will ever forgive him.

But, *An Inconvenient Truth* by Al Gore hasn't yet had its moment of truth. Many, if not most, still believe Gore's story and that the future of civilization depends on an immediate and massive governmental response. This response is being seriously considered, in spite of the fact that Gore's movie had many significant errors. And whether or not you believe in man-made climate change, all of our efforts will have a negligible effect on the climate, and, yet, probably a major negative effect on our economy.

Again, our subconscious works very hard to filter out any data that would conflict with what we strongly believe. Many in power know this, and use it to their personal advantage.

I discuss Gore's movie in greater depth in Chapter 4.

Here is a personal example of how the power of stories can change people in significant ways:

During the first week of January in 1999, my dad and my former wife Kathie were standing at a corner in Sunnyvale, California waiting for the light to change. When it did, they stepped into the crosswalk and started to cross the street. Then they noticed a car making a left turn and coming right at them ... they stopped and backed up ... the car stopped ... my dad went ... the car went. And a terrible accident happened.

My dad was very seriously injured, with numerous bones broken in his pelvis and leg. He was in a drug-induced coma for ten days, only to awaken to massive pain. His chances for recovery, at eighty-three years old, weren't very good. He spent about a month in the Stanford Hospital and then was flown to a hospital in Billings, Montana, the city where Mom and Dad lived most of the year.

While we were all very concerned about his body's ability to heal from such a major injury, we were just as concerned about his attitude. When I called him in the

The Power of the Story

hospital, I was always saddened to hear his voice. He was difficult to understand, and all of his usual zest for life was missing. We were afraid the chances of his getting well, if he didn't have the will to do so, were not very good.

I was VP of product development for a startup in Austin, Texas, at the time. One day in late February we were in a company planning session with an important advisor. The first thing he did was check with each of us to see if we were going to be ready to go to market by September. Marketing reported: ready. Sales reported: ready.

Then he looked at me and said, "How about Product Development? Are you going to be ready with the product?"

There was a moment of silence and then I replied, "There's no way we're going to be ready. We don't have a design specification for a product, only a visionary concept. In addition, we haven't even begun to hire our technical staff; you can't do a realistic technical design or schedule without input from the people who will be responsible for delivering the product."

This didn't go over well. In fact, the advisor looked up at us and said, "Let me know when this group's got its act together." Then he left the room.

With fire in his eyes, the sales VP said, "I've been working seven months to help build a successful business, and you destroyed everything in a couple of minutes!"

Everyone else just looked sad. It was one of the worst days of my working life.

When I got home I told Barbara, "It isn't worth it. I'm fifty-eight and I've already been there and done it. I don't want to put up with this arrogant ignorance anymore.

I think I'm going to give my notice." Then I told her what had happened.

Barbara was supportive and encouraging, but I still needed to vent. I called my son, "Jon, I need to talk ... have you got the time to listen?"

"Sure, Dad. What's the problem?" I told him my story. After I was through, he said, "Doesn't sound like my dad."

"What do you mean?"

"You know, Dad, being a minister isn't all peaches and cream. There are times when I just want to chuck the whole thing, quit worrying about all the politics, tell some of the people to go take a hike. You know what I say to myself when I feel that way?"

It seemed to be a rhetorical question, so I waited for him to continue.

"I say, 'What would Dad do? Dad would look at me and say, 'When the going gets tough, Freggers get tougher!'"

I hate it when my kids throw my own words back at me like that! When we finished the conversation, I looked over at Barbara and said, "Jon talked me out of resigning ... they can fire me, but I won't quit. And I'll keep doing and saying what I know is right."

I took a deep breath, was silent for a moment, and then it hit me ... <u>this was the exact story I needed to tell Dad!</u> He was having a tough time. The solution: "When the going gets tough, Freggers get tougher!"

Better yet, it was a real story, one I could easily share with honesty and passion.

The Power of the Story

I called him immediately. He answered with the same hard-to-understand, lackluster voice that saddened me so. But as usual, he wanted to know, "How's your day been?"

"Probably the worst day of my working life," I said.

"What happened?" he asked with fatherly concern.

So I told him. At the end I said, "So you know what happened when I got home?"

"What?"

"I called Jon and told him my story. Do you know what he said? He said, 'You know what I say to myself when I feel that way? I say, 'What would Dad do?' Dad would look at me and say, 'When the going gets tough, Freggers get tougher!' "

There was a moment of silence, and then my dad asked, "What are you going to do?"

"I'm going to stick it out. They can fire me, but I'll be damned if I'm going to resign. And I'm not going to stop telling them what I think they need to hear either."

At that point the conversation turned to other things.

A few days later I called him and got the surprise of my life ... it was my old dad back again, his voice full of humor and that zest for life I have always loved.

"I can't talk for long, Brad. They're going to be here in a minute to take me to therapy, and I've got to get ready. It looks like I'll be going home in a couple of days, and there's no way I'm letting your mother take me to the bathroom."

During the entire time of my dad's illness, my brother Dennis had taken responsibility for keeping family and friends up-to-date on his progress. He did this with email messages; here's the final message that he sent to all those people who cared so very much.

Dennis' Message

We arrived in Billings, Montana, on April 19th, my mother's 80th birthday. It was about 6 p.m. when I drove into my folk's garage. Mom opened the door from the door immediately and greeted us with a big smile. As I stood at the kitchen door and looked into the living room, I was amazed to see my father walking towards me. His big smile belied his recent past, as did his limp-less gait. He carried a cane, just in case, but didn't need it!

HE LOOKED EXACTLY AS HE HAD BEFORE THAT EARLY JANUARY DAY! We reveled in the wonder of it. I've begun to call him the "Miracle Man." He's not up to his regular two to four miles a day yet, but he's walking about a half to one mile presently. The transformation from when we last saw him is simply mind-boggling!

I'm convinced much of his improvement can be placed squarely on all of you beautiful people. Yeah, I know genetics was a great help, too; but all your cards, notes, and e-mails simply blew Dad away! More than once he said, "I didn't know so many people cared!"

Thank you so much everyone! I'm so happy this last installment of my father's saga was full of joy.

I will always wish the same for you.

Love, Dennis

The Power of the Story

3
A Solution for the Energy Crisis

There is one reasonable solution that would solve our energy crisis easier and faster than any other proposal being offered. It is a solution that will take commitment and sacrifice. As the commercial said, "You can pay me now, or you can pay me later."

Here are the major components of our energy crisis:

- Dependence on foreign oil to run our economy, most of which is being used to power our automobiles;
- World-wide pollution caused mainly by automobiles and trucks;
- A move by environmental lobbies to limit individual freedoms in order to "save" the world from catastrophic climate change;
- A dramatic increase in individual and corporate taxes to fund the changes needed to "fix" the societal problems that have "caused" global warming,
- At least three countries that would like nothing better than to see the United States defeated—and, if they can't beat us militarily, they will do everything in their power to beat us economically. These countries are Russia, Iran, and Venezuela.

A Solution for the Energy Crisis

Sooner or later we are going to have to face these issues; the reasonable solution offered in this chapter allows us to move forward immediately.

Interestingly, a public announcement of our commitment to this solution would have the immediate effect of lowering the world price of oil, with the result that gasoline prices would quickly drop back to more acceptable levels; and, Russia, Iran, and Venezuela would have less cash with which to cause us, and the rest of the world, problems.

Lowering the price of oil would also lower the price of transportation, food, and other commodities. The increase in price of these items, caused in part by the increase in the price of oil and other decisions made to try to "stop global warming," is already causing problems worldwide with food riots in Mexico, Egypt, and other countries.

So, what follows is a reasonable solution to the energy crisis that would have additional benefits that could help people throughout the world and assure our societies future social-economic health; as well as cause major difficulties to those who would like nothing better than to see the United States as a minor player on the world's stage.

A Reasonable Solution to the Energy Crisis

I am a loyal, committed citizen of the United States who believes that the legislative and executive branches of our government have let us down by catering to the many powerful special interest groups that control our elections; in this case, big business and the environmental lobby. Because of the power these lobbies exert over our government, it is almost impossible to make the right decisions regarding our ability to achieve energy independence and our future as a nation. The personal interests of big business and the environmental lobby essentially block any legislation that would solve this problem.

The following statements are the basis for this chapter and the solution I am presenting.

- It would be to the world's advantage to minimize pollution of any kind.
- It would be to our advantage to become energy independent.
- It would be to our nation's advantage to limit the influence of Russia, Iran, and Venezuela in regard to the world's oil supply.
- It would be to our advantage to be proactive in solving the energy crisis.

- There is no need to limit the personal use of gasoline, light bulbs, or any other commodity. Given the proper choices, the market will handle the issue.
- We must be very careful about what laws and, therefore, requirements and restrictions we allow government to place on our citizens, organizations, and/or companies.
- It would be to our advantage to hold a vision of the future where we have succeeded in becoming energy independent and continue to be the country that people throughout the world look up to and dream of becoming citizens of, a model that despots fear.

There is no doubt in my mind that my proposed solution is the best for many reasons. However, it is not a perfect solution and the interests of the two major lobbies I have mentioned will be impacted negatively. In addition, there will be major problems for local, state, and federal governments with the resulting loss in tax revenue.

    This means that the rest of us will have to apply sufficient pressure on our elected representatives to force them to create legislation that will create a path to energy independence. This proactive legislation will provide the vision needed to keep us on the path to social-economic health and success.

By the way, if you believe that a strong United States is a threat to the world, you might as well stop reading.

History has shown that the United States is the most caring, compassionate, and sharing nation the world has ever seen.

Never before has a conqueror (World Wars I and II) been so kind to those conquered. Never before has a nation with ultimate power (the atom bomb) chosen not to use it to control the rest of the nations of the world. Never before has a conquering nation allowed those conquered to form their own governments and make their own alliances, whether or not the conquering nation agreed with the decisions made. Never before has a nation been so willing to help those in need, no matter who they were, or what God they prayed to.

Of course, we aren't perfect, but we are the best … and I, for one, want that to continue to be true.

So, how do we become energy independent, in a reasonable length of time, while minimizing pollution, especially carbon-based pollution, and creating an environment where individual choice is a commitment, business can thrive, and our nation can look forward to decades of socio-economic health; while, at the same time, limiting the influence of Russia, Iran, and Venezuela? Seems like a tall order, doesn't it? Actually, it is a relatively easy problem to solve, with the clear understanding that any

A Solution for the Energy Crisis

change, no matter how positive, causes problems for some.

There is no doubt that transportation vehicles (cars and trucks) are the greatest polluters that exist in the world today. And, this problem is not going away; in fact, with the increase in the use of automobiles throughout the world, especially China and India, the problem is only going to get worse. If we could just solve the vehicular pollution problem, we would go a long way to solving the pollution problem world-wide. In fact, solving this one problem would have a greater positive effect on the environment than any other solutions I have seen proposed.

Any solution that suggests doing without automobiles will never work; it is naïve to even consider such a solution. Likewise, it is unreasonable to consider limits to socio-economic growth. In fact, it is dangerous to try to implement such solutions, as the impact on our economy would likely result in a depression that would cause greater suffering than global warming will ever cause.

First of all, let's take a look at our alternatives to gasoline as vehicular fuel.

Propane, Natural Gas, and Methanol

Propane, natural gas, and methanol are popular alternative fuels that are readily available. When compared with gasoline, these fuels produce fewer air pollutants and greenhouse gas emissions, but, they are not a zero-pollution alternative. Existing vehicles can be converted to use these fuels, or a combination of one of them and gasoline; and there are some factory-produced vehicles being manufactured. While the use of these fuels continues to increase world-wide, there are major problems with considering these fuels as a replacement for gasoline.

First, natural gas is not a renewable resource and while currently readily available, it has many other important uses, which makes the decision to use it problematic as a major source vehicular fuel.

Finally, there is an infrastructure issue that goes beyond the manufacturing of vehicles that use these fuels. The issue is building the service stations that would be needed to service the hundreds of millions of automobiles in the United States; and then, manufacturing and transporting the fuel to the millions of service stations needed. And, of course, natural gas and propane are both extremely flammable; while much has been done to minimize the danger, the potential for loss of life due to gas explosions is still there.

These fuels provide an alternative in specific cases (fleet vehicles, lift trucks, etc.) but this is not a good solution as the major replacement fuel for gasoline.

Ethanol

Ethanol can be made from renewable resources such as corn, sugar cane, grain, wood, etc. Most conventional vehicles can use up to 10 percent ethanol without any modifications. Many states have filling stations offering such blends. Use of ethanol reduces greenhouse gas emissions. These are the positives; however, there are many negatives to using ethanol as a replacement for gasoline.

The major issue here is corn ethanol. Corn is a food, and not just any food. Corn oil is used in millions of products; our livestock (cows, pigs, chickens, etc.) eat corn; and then, of course, are the hundreds of millions of people around the world who depend on corn as a major source of nutrition.

Many, who profess to care about the world and the people in it, seem unaware of the impact that the use of corn ethanol has already had on the world wide food supply. The price for all corn-fed livestock has increased significantly, driving up the price of meat, eggs, and dairy products, as has the price for all corn products, including the staples in some countries, especially Mexico. And, this is just the beginning. If all the available land in the U.S. were set aside to grow corn, it would still only supply a small percentage of our national needs.

Corn is expensive to convert to fuel and, unlike oil, pipelines cannot be used to take it from manufacturing to distribution. Ethanol must be trucked to all major distribution centers and then trucked to the service stations.

For all of these reasons, there is a strong likelihood that corn-ethanol fuel will be more expensive than gasoline. In other words, converting nationally to corn ethanol to solve the energy crisis is a very bad decision, and the only reason it is being suggested is that the technology exists and it looks like it can have an immediate impact on the problem. This is extremely short-sighted thinking, thinking that is, as I have already stated, causing problems world-wide.

The other ethanol choices aren't much better. They still take lots of land to grow the source, which means cutting down virgin forests and limiting the use of the source for other uses ... thereby driving up the prices of other commodities. There is already great concern about the loss of our virgin forested land, especially in the Amazon basin; converting to ethanol would require that these forested lands be used to grow the plants needed to produce ethanol.

Even the use of scrap cellulose, corn husks, wood scraps, etc. would have an insignificant impact when one considers the amount of fuel needed to run the world's vehicles today and in the future.

Biodiesel

Biodiesel can be made from vegetable oils and animal fats. It produces fewer emissions than regular diesel and is biodegradable. However, the biggest problem here is the short supply. There will never be enough of this stuff to even begin to replace even a small percentage of the fuel needed.

Hydrogen

Hydrogen is often touted to be the vehicle fuel of the future. There is no doubt that when produced from renewable sources, hydrogen has the potential to be one of the cleanest alternative fuels. Hydrogen gas is the most abundant element on earth, but it needs to be extracted from compounds such as natural gas or water before it is available as a fuel. It will likely be many years before hydrogen is a cost-effective and commercially available fuel. In other words, the technology to do this at a reasonable cost, while having a negligible impact on the environment, does not yet exist. Some experts estimate that it might be as long as 25 years before hydrogen is ready to replace gasoline as a vehicular fuel.

And, once we've got the technology to produce it, we still have to deliver it to the consumer. Again, the infrastructure issue raises its ugly head. Creating an entirely new distribution channel will take billions of dollars, large resources, and assure us that the cost of the fuel will most likely be much more expensive than gasoline. And, hydrogen fuel is also highly explosive with very real danger during transportation and even in the home garage. All of this means that hydrogen, as a competitor to gasoline, is a long, long way into the future.

Electricity

Electric-powered vehicles produce no tailpipe emissions; however, they do use electricity from sources such as electric power generating facilities which do produce pollution. However, there have been, especially here in the U.S., significant advances in controlling the pollution from these facilities. There is no doubt that it is much easier to control the pollution from an identified number of plants, than it is to control the pollution from 300-million plus vehicles.

The major problem with electric-powered vehicles has been the distances that they can travel without having to recharge the batteries; however, technologies continue to improve, with the Tesla all-electric car getting a range of 220 miles on a single charge (3.5 hours) with a 135 mpg equivalent to a gasoline-powered car (about 2 cents a mile). The battery life is approximately 100,000 miles. Regarding other performance issues, the Tesla accelerates from 0 to 60 in less than 4 seconds and has a top speed of 125 miles per hour.

This performance pretty much eliminates all other options, which, in addition to poorer performance, are more expensive than gasoline to operate, while the all-electric costs significantly less than gasoline to operate. I haven't even mentioned maintenance, which is another advantage of the all-electric. The main maintenance issue with an all-electric is the need to rotate the tires and replace the brakes; there are no oil changes or tune ups needed. Add all of this to the prior-mentioned facts that we know how to build an all-electric, and we know how to generate electricity, and it seems like an obvious solution. … But, how about the infrastructure issue?

When I'm talking to groups about this option, I ask the question, "We have service stations to refuel our gasoline-driven vehicles, so … how many service stations exist to refuel our electric-driven vehicles?" Usually the response is well under 1000.

Almost everyone is surprised when I tell them that there are well over 300-million refueling stations already in existence. ... This includes, of course, every home and business in the U.S. The infrastructure for refueling electric vehicles already exists; you just pull your vehicle into the garage and plug it in. And, even more impressive, adding more service stations at every parking garage, motel, office, etc. will be a very simple process.

Even after mentioning all of these positives, I still get objections. One of the major objections I get is, "Where is all this electricity coming from?" First of all, in the beginning, this will not be much of an issue. Remember, most of this recharging will take place overnight, when our electrical usage nationwide is at its lowest levels. Additionally, individuals have the capability to create their own electricity; some are doing that currently, using either solar cells or windmills.

There is no doubt that, initially, we will have to build additional electric generating power plants. However, as I stated earlier, it is much easier to control pollution from a number of power plants than from 300-million plus vehicles. What is important here is that we make reasonable decisions, which means we have to ignore the fanatical side of the environmental lobby—currently the major block to achieving the resources we need.

A Solution for the Energy Crisis

We must keep in mind that, with this focus, we would be on the road to dramatically reducing vehicle-caused pollution in a very short time and, effectively, eliminating it altogether in the long run. Transportation vehicles account for about two-thirds of all pollution and possibly one-third of carbon-based pollution (depending on what expert you're using). Just solving half of this problem would provide better results than most other, much more expensive solutions, can even dream of.

Regulations on coal and oil burning, and nuclear power plants must be kept at a reasonable level, while we continue to do the research needed to improve our ability to eliminate the pollutants these plants tend to produce. At the same time, we must continue to improve the efficiency of solar, wind, and tidal-generated electricity, all of which have essentially zero emissions. This research is continuing at a rapid rate; for example, research in solar cell technology is advancing at an exponential rate. It is conceivable that we will see extremely efficient solar cell technology within the next decade.

If you are wondering who would purchase an all-electric vehicle, just watch the documentary titled, "Who Killed the Electric Car?" The movie is currently available on YouTube in one form or another, and I am sure that it can be purchased from your local bookstore or an online retailer.

The mistake most experts make when considering the viability of a car that takes 3.5 hours to fill up after driving about 200 miles, is forgetting completely the number of people who would love to own a vehicle designed only for trips within 100 miles of their home. Millions drive less than 200 miles roundtrip every day, spending the vast majority of their gas money on relatively short trips, not on long trips.

There is a strong possibility that families will either have two cars, one for everyday driving and one for long, vacation-type trips; or that they will buy a car for everyday driving and rent when they want to take a long trip. "Who Killed the Electric Car?" strongly supports the number of people who would love to have an efficient, well-performing, electric car, even if the range was significantly less than 200 miles.

And this solution is only temporary. There's no reason not to expect that within the next decade we will have batteries that enable the all-electric to go 500 miles on a single charge. And, then, I can easily envision a 1000-mile battery in the future, or, even better, an efficient solar cell that will enable a vehicle to automatically recharge. There is no doubt that the all-electric is the vehicle of the future—there is no other alternative that can compete with it.

A Solution for the Energy Crisis

The desirability of the all-electric was proven in California in the 1990s. At that time, the Air Resources Board (ARB) required that in 1998, 2 percent of the vehicles produced for sale in California had to be Zero-Emissions Vehicles (ZEVs), increasing to 5 percent in 2001, and 10 percent in 2003. GM and a few other major auto manufacturers produced all-electric cars to meet the requirements, and the lucky few that were able to lease the GM vehicle gave it an outstanding review. This was achieved in spite of the fact that the GM all-electric's range was between 50 and 100 miles.

In fact, the car was so successful that GM (and the U.S. Government) convinced California to eliminate its requirement for a zero-emission automobile. As soon as the requirement was lifted, GM immediately took back all of their leased electric cars and destroyed them. If any remain, they are kept under lock and key and only GM executives are allowed to see them. The reasons (I believe they took such drastic action) are discussed when I cover the negative implications of the all-electric vehicle.

It is interesting to conjecture about what took place at GM when they realized that they had to produce a zero-emission car. My guess, and it is only a guess, is that initially they didn't think it was possible. But, they needed to make an honest attempt before they complained to California and the Federal Government. So they formed a small "skunk works" and gave them the challenge, including the money, resources, and authority to make it happen.

I'm sure that the GM executives were as surprised as anyone when the car that the skunk works developed was so successful. Since they couldn't allow this to happen, they called on the Federal Government and pressure was applied on California to change its requirement … thus, it was GM, the oil companies, and the Federal Government that "killed the electric car." As you will see, they had, as far as they were concerned, very good reasons to take this action.

Many also ask me about the hybrid. There is no doubt in my mind that most hybrids are purchased by people who want to show that they care about the environment. While the gas mileage in city driving is greater, there is little improvement in freeway driving, especially long distances. In addition, the issues of dealing with a car that has both a gasoline and an electric engine is hardly worth the effort, considering the amount of energy and pollution saved. Another reason for the marketing of the hybrid is that it is an electric car that still needs a lot of maintenance, which is significant to the profitability of the auto industry.

A Solution for the Energy Crisis

So far it seems like the all-electric car is a "no brainer" for the future of vehicular traffic in the U.S. (actually the world) for these reasons:

- Current equivalent gas mileage is 135 mpg (10 times some trucks and SUVs, and well over twice the best of any of the current hybrids), and that will only get better as battery and solar cell technology improves;
- Technology to manufacture the car and produce the fuel exists and has been fully tested;
- Performance is equal to, or better than, gasoline-fueled cars;
- Maintenance costs are way below that of gasoline, or any other alternative;
- Infrastructure to deliver the fuel is already in existence and is actually superior to what exists for gasoline.

And, I haven't even mentioned how silently they run. ... Yes, we also get an end to noise pollution. So, if all of this is true, what's the problem?

The negatives of the all-electric are essentially economic, with the fanatical fringe of the environmental lobby adding frosting to the cake. It is interesting that these two lobbies, which have no great love for each other, form such a strong alliance against us making the right decisions for energy independence, our socio-economic health, and the future of our country.

The profitability of the oil and auto companies currently depends on the continued use of gasoline for vehicular fuel. For the auto companies, it's the servicing of these vehicles; and for the oil companies, it's the importance of gasoline. Any decision supporting all-electric vehicles as a replacement for gasoline-fueled vehicles would significantly impact the profitability of these companies.

There are also strong legal precedents against making any corporate decisions that would be negative to shareholder's investments. This means that these companies are duty bound to fight any decision that would move us from gasoline. And, don't forget, the vast majority of the stock of both the auto and oil companies is owned by average U.S. citizens in either stocks, pension plans, or mutual funds.

In addition, nine of the ten largest companies in the world are either oil or auto companies. Any significant change to their profitability could potentially cause a world-wide depression that could result in millions suffering great hardship. I believe that this was a major reason for the government siding with GM to convince California to eliminate the zero-emission requirement. And, let's not forget the importance of gasoline tax revenue for our local, state, and federal governments. All things considered, it is no wonder that the pressure to keep things as they are is immense.

A Solution for the Energy Crisis

I'm less worried about the success of the all-electric car causing a world-wide depression than I was ten years ago. Recently we have seen that large companies are able to survive major upheavals in their market. The best example of this is Kodak which experienced the total destruction of their market when digital technology replaced film in a remarkably short period of time. After a major reorganization, Kodak is again seeing profits rise, finishing 2007 with a very nice increase.

This doesn't mean that it will be easy for oil companies to replace their revenues, but they will have time to make the plans necessary for their continued success. There is no doubt that they will find ways to use oil to produce the additional electricity that this solution demands; and, that they will quickly find ways to control the emissions from those plants.

The auto companies will have an easier time of it, since the bulk of their revenue comes from the sales of cars. However, there are millions of people dependent on the fueling and servicing of gasoline-driven automobiles; they will, of course, have to make adjustments as more and more of us move from gasoline vehicles to the all-electric vehicles of the future. And, I'm sure government will find a way to replace the lost gasoline tax revenue.

As you can see, these are powerful reasons for both the government and the oil and auto companies to fight any major switch to the all-electric vehicle. In fact, ten years ago I would have supported this decision. ... However, times have changed, and we must revisit the need to go all-electric.

Below are the reasons we must seriously consider making the tough decisions necessary if we are going to continue to be a major player in the world, instead of a "has been," as we watch the rest of the world overtake us and control not only our future but our daily lives.

- Our entire infrastructure is dependent on the energy we purchase from other countries, who could just as easily decide to begin selling their oil to others.
- Our safety as a nation is threatened by three countries (Russia, Iran, and Venezuela) that, because of the amount of oil they control, have the funds and influence necessary to cause us harm; potentially great harm, if we don't act soon to solve this problem.
- The world is very concerned about the impact of gasoline-fueled vehicles on world-wide pollution and climate change.
- Our nation's vision of the future is becoming increasingly negative as the economy seems to be moving toward a full-blown recession, and we don't seem to be able to do anything proactive to change the direction things are going. We're a nation of doers, we don't like sitting around waiting for things to happen; we like being in control.

A Solution for the Energy Crisis

All of these issues are solved by a concentrated focus on the all-electric vehicle. Here's what we need to do:

- As a nation, we should commit to all-electric technology, including a "Manhattan Project" to develop, first, 500-mile range, and then, 1000-mile range electric vehicles that meet at least the current performance standards of the Tesla all-electric vehicle. At the heart of this project will be the research and development needed to mass produce the battery that will enable an all-electric vehicle to have this range.
- We must support, in every way possible, the development of extremely efficient solar cell technology. Achieving a efficient way to create electricity directly from sunlight will be a boon to all of humanity.
- Congress should immediately pass a zero-emission requirement that mandates that a minimum of 10 percent of all vehicles manufactured be zero emission by 2020, with the initial zero-emission vehicles being available to the public by 2012—or earlier if possible. (Remember, the technology exists and has been proven.)
- No laws should be passed limiting the ability of individuals to purchase gasoline-powered vehicles. There is no need to limit freedom in this regard. In fact, there should be no need to limit individual freedoms at all in regard to the issue of climate change and/or global warming; all that is needed is reasonable governmental policy.

- Government funding should be immediately available for the mass manufacturing of the batteries, etc. needed for the initial zero-emission vehicles so that they have a minimum range of 200 miles. Or, a tax credit could be given to individuals purchasing all-electric vehicles, to help offset the current high cost of efficient batteries.
- Congress should immediately pass reasonable requirements for the building of electric power generating facilities using either coal, oil, or nuclear power. There is no doubt that our society will need much more energy in the future; we need to start now if we are going to get the job done in time for our future needs.
- Congress should immediately offer significant incentives for the development of all cost-effective, alternative methods of generating electric power.

The "Manhattan Project" to develop 500- and 1000-mile range electric vehicles with at least the performance standards of the Tesla all-electric has advantages beyond the boon it would be to our goal of energy independence. Actually, calling it a "Manhattan Project" is probably a misnomer. The Manhattan Project was highly secretive and involved the development of the Atom Bomb. Einstein proposed the project after hearing that the Germans were trying to do the same thing. He knew if Hitler got the atom bomb first, that he would use it to conquer the world and wouldn't care how many were killed in the process. Likewise, he believed that the U.S. would not do that.

However, it is important that the project to develop 500- to 1000-mile range electric vehicles not be kept secret for a couple of very important reasons. First, this commitment should have the same effect on our society that Kennedy's pronouncement for a landing on the Moon before the end of the century did. This goal was extremely positive in regard to our nation's socio-economic health.

For those of you who may believe the money was wasted, the initial research that I did for my Master's thesis suggested that the government received upwards of $5.00 in taxes for every dollar spent on the Moon landing, kind of a hyper "trickle-down" effect. If this is true, the goal to reach the Moon by the end of the century was one of the best investments our government has ever made. This kind of positive vision of the future is essential for the socio-economic health of a society. I will cover the second reason why this project must be announced to the world at the end of this chapter.

While I am against government involvement in private matters, there are times when it is essential. For example, in our society it is against the law for a man to beat his wife and/or molest his children. In the instance of mandating zero-emission requirements, it is critical that they be stipulated by Congress; the auto companies cannot make this decision on their own. If they did, there would be no end to the lawsuits being brought by the companies' shareholders. Likewise, Congress has to hear from a vast majority of U.S. citizens that this plan is essential. There is no doubt that the oil lobby will also fight it with every resource they have available.

However, I want to make note that the requirement is designed to get zero-emission vehicles on the road; there is no governmental requirement beyond the initial 10 percent. There is no doubt in my mind that the market will take care of the rest. Once you can buy a car that will handle all of your everyday needs, never need to go to a gas station for a fill-up, and cost about 2 cents a mile to operate, all of this with excellent performance … the problem will quickly be, meeting the demand.

Owning a gasoline-powered vehicle should never be against the law. Hobbyists and traditionalists should be able to have their cars … although … there may come a time when they will have difficulty finding a gas station.

Congress will also have to be shielded from the fanatical end of the environmental lobby. They will do everything in their power to stop the construction of nuclear and oil- and coal-burning power plants, regardless of the safeguards put in place. Without the additional power that will be needed, this plan will fail and we will find ourselves no better off, still struggling to survive in an evermore competing world; a world that will care less and less about what happens to us. In fact, some will celebrate our decline.

Finally, regarding Russia, Iran, and Venezuela, there's no doubt that these three countries are using their oil revenue to increase their influence around the world and build up their militaries. Without the windfall profits these countries are getting from their oil, they will have a much harder time gaining the influence and military strength they need to achieve their goals, which I assume include replacing the United States as the world's major economic power. So, if we could dramatically lower the revenue they are getting from their oil, we would significantly impact their ability to realize their future plans.

    It is critical to understand that we are in the middle of an economic world war. And, in this instance, the competition is not interested in compromise; they are out to get us and will celebrate in the streets if they are able to make it happen. There is a major rule in negotiation: when the other person/company/country is only interested in their winning and your losing, you will never win by compromise or collaboration. In this instance, you, also, must commit to winning; all the diplomacy in the world will not change the situation, but only delay the inevitable. So how do we win? By lowering the price of a barrel of oil to half of what it is right now; a price that reflects the supply and demand issue much more accurately. The process for accomplishing this is relatively easy and will happen automatically if this proposal is acted on.

While the price of the barrel of oil is influenced by many factors, I believe most experts would agree that the oil speculators are the major reason. Speculators drive up the price by anticipating the future and determining what a barrel of oil will cost a year or more from now. If they decide that the world situation will continue to be volatile and that the United States will continue to do nothing to change the situation, they will determine that the price of oil will continue to increase and, therefore, they will continue to bid it higher and higher. This type of investment is very much influenced by what they *believe* will happen. So, the answer is to create a future scenario where the price of oil will plummet.

This happens in business all of the time. It is surprising to me that Congress seems to have no idea that, in the main, they are the problem. By refusing to do anything positive toward energy independence, they are creating the situation that will assure continued increases in the price of gasoline.

Microsoft and other major companies understand the power of limiting the expectations of investors.

A Solution for the Energy Crisis

If Microsoft is concerned that a new company will develop a technology that might impact their future profitability, they announce development of a similar technology. Just the announcement of their intention to develop this technology scares investors away, and the new company finds that their investor base has dried up. There are no more skittish individuals, than investors … create a scenario where they are likely to lose their investment, and they will run for the hills.

So, once Congress announces the "All-Electric Project" to develop a 500- to 1000-mile battery and the requirement for an available zero-emission vehicle by 2012 at the latest, the price of a barrel of oil will plummet to, probably, half of what it is now within six months. This will dramatically impact Russia's, Iran's, and Venezuela's future plans … and, additionally, ensures us of a reasonable price for gasoline as the change is taking place.

This will, of course, be our biggest challenge: once the price of gasoline gets back to $2.00 a gallon, we must stay committed to our goal of energy independence that can be achieved through the development of effective and efficient all-electric vehicles. It is critical that we stay committed to this course of action. Our future and the future of our children's children depend on it.

4
What's New about Climate Change?

Why are we so concerned about climate change (global warming)? In my opinion, the public, and therefore politicians, are concerned mostly because of Al Gore's movie, *An Inconvenient Truth*, and the infatuation of the mass media (TV, Internet, and newspapers) with the theme of the movie, which is right up their alley—the sky is falling and it's our fault.

Chapter 3 discussed a reasonable solution for the energy crisis. In that chapter I mentioned two issues that could be shelved if the proposed solution were implemented. However, that is probably wishful thinking; the issue of humanity-caused climate change is a "faith-based" belief, not one based on reliable scientific evidence. I'll get more into this in a moment; however, let me remind you of the two issues.

- A move by environmental lobbies to limit individual freedoms in order to "save" the world from catastrophic climate change.
- A dramatic increase in individual and corporate taxes to fund the changes needed to "fix" the societal problems that are causing climate change.

What's New about Climate Change?

It should be noted that the operant phrase now is "climate change"; if you remember, not that long ago it was "global warming." It became necessary to change the phrase, or spin, when it was discovered that the world had quit warming, and that, in fact, we were probably in-store for some pretty cold weather. By changing to "climate change," the activists and other special interest groups could still blame humanity (read that "western civilization," especially the United States) for the problem, ignoring completely that the world's climate is continually changing—and has been since the beginning of climate—whenever that was.

I am very concerned about governments getting into the climate change business even more so than they are now. (Governments are the largest funders of climate change research and focuses those funds on scientists that support the claim that western civilization is the major cause of global warming).

Our government has shown us time and time again that their approach to problems, especially complex problems, is almost always the wrong approach. Even worse, once bad decisions are implemented and bad laws passed, it is extremely difficult for the government to admit its made a mistake and correct it.

Let's take a quick look at two of the more recent "wrong" decisions made by our government; both of which were made for the express purpose of limiting humanity's impact on climate change and those presumed to be affected by it.

- The law mandating the use of corn-ethanol gasoline blends as a replacement fuel for gasoline.
- The law putting the polar bear on the endangered species list.

Both of these laws are the result of an ignorant group of politicians listening to environmental activists and others with less altruistic motives. This truth of this will be clearly shown in this chapter.

Mandating the use corn ethanol in gasoline/ethanol blends was a terrible decision. This is something that the vast majority of experts agree on; even the media (*Time* and *Newsweek*) agree. Yet the U.S. government is doing nothing to change the mandate. Iowa's state government is actually debating increasing the percentage of ethanol to 15 percent.

It is decisions like this that support the contention that Congress, while probably containing a number of intelligent individuals, finds it very difficult to make reasonable decisions. Instead of being the leaders that we need and want, they end up being followers, continually testing the winds of popular sentiment and listening to "experts" with personal agendas—who will ultimately supply them with the funds they need to get re-elected.

I've covered this issue in some depth in the previous chapter, so I see no need to go into it in depth at this time. However, I do want to reiterate that there are no sensible reasons for mandating the use of corn ethanol gasoline blends as a replacement for gasoline.

Now, let's take a look at the polar bear and the lack of reasonable decision making that led in May, 2008 to their being placed on the endangered species list as "threatened" (not quite endangered but heading there).

First, and most important, the polar bear is not currently an endangered species, nor are there any signs that they are "heading there"; all research, including Eskimos' observations, support that they are a very robust species in no immediate danger or threat of extinction. In fact, polar bear populations have grown from 5000 in 1940 to over 25,000 today.

The two issues that one must accept before we can consider the polar bears under any kind of threat because of climate change or global warming are:

- Polar ice is melting because of mankind-caused global warming;
- Polar bears need a sufficient amount of polar ice to survive.

Regardless of whether there is sufficient data to show that global warming is being caused by the amount of CO_2 humanity is releasing into the environment (which is still hotly debated), there is new data that shows, fairly accurately, that the currently "rapid" melting of arctic ice is really due to a temporary change in the wind direction and has nothing to do with climate change. These two issues are discussed in much greater depth in an article posted on the American Enterprise Institute's website, "Is the Polar Bear Endangered, or Just Conveniently Charismatic?" by Kenneth P. Green.

The article also discusses the supposition that polar bears need arctic ice to survive. It seems obvious that this contention is extremely weak, one that could only be believed by very ignorant people.

The reason that I state this so strongly, is that we have absolute proof that the arctic region has seen much greater warming in the past and that the polar bear has been up there for well over 100,000 years. Ice cores have shown dramatic changes in climate in the area over those many millennia, yet the polar bears survived them all.

What's New about Climate Change?

Therefore, the belief that polar bears are endangered because of global warming is truly an ignorant belief … or, one held by those, with their own agendas, who are using the polar bear and spinning its story, in order to accomplish something else much more critical to them, but dangerous to the wellbeing of the rest of us … the elimination of any oil and gas drilling by the U.S. throughout the arctic—probably throughout the U.S.

This desire on the part of the environmentalist, whose current high priest is Al Gore, to make it impossible for the United States to become energy independent, is puzzling. If they are serious about this issue, then they have been badly misled. There's no way that we, the world's people, will not suffer more from the regulations they want to put into place, than we ever will from the amount of global warming we will be able to prevent from happening.

In my opinion, Al Gore is a faux-environmentalist who figured out a long time ago that he could make himself a fortune if he could convince certain key people and the media that our selfishness as a society has caused the world to enter into a phase of global warming that will kill and/or displace millions and millions of people.

According to Gore, and his movie, *An Inconvenient Truth*, the main problem is the CO_2 that we are pumping into the atmosphere. However, he says, it isn't too late, all it takes is for the government to increase taxes by billions of dollars, and use that money to fix the problem.

I will say this—he definitely achieved his goal! The entire world is on the global-warming bandwagon (what a waste of the Nobel Peace Prize). I'm not sure what will happen when enough people realize that they've been hoodwinked. However, I am sure that his legacy is set; he will go down in history as a fraud, not the savior he pretends to be.

I've been using some religious terminology, because that is what environmentalism is, a religion. The environmentalists' god is Mother Earth and their chief priest is Al Gore. Doing anything to cause harm to Mother Earth is a sin of the first magnitude, and western civilization has been tried and convicted of this sin. We must pay a penalty if we expect to get back in her good graces. The penalty is: higher taxes and a simpler lifestyle (Al Gore seems to be exempt here).

Like religion, environmentalism, expects complete devotion from its followers; nothing voiced by the high priest can be challenged—those who dare are shunned by the true believers as heretics (global-warming deniers)—intelligent questioning, dissent, is not allowed. Environmentalism, just like other religions, has its superstitions and rituals. You bring your own (usually green) reusable bags to the grocery store, you change all of your light bulbs to the new fluorescent variety, you spread the word so that others will learn the "truth" and be moved (convinced) to join the ranks of the true believers.

This religion is so critical that all of humanity must join its ranks; anyone who doesn't risks the dangers of becoming a social outcast.

As you can tell, I'm angry. I want to toss this faux religion "under the bus." Don't get me wrong, I think intelligent conservation is a good thing. I drive a Scion XA, considered to be one of the most environmentally friendly cars sold. I just purchased an electric lawn mower so I wouldn't foul the environment with both chemical and noise pollutants, and, we take a bag to the grocery store, too. I'm all for saving a few trees if I can.

All of that, and more, we do by choice, not by mandate. I'm angry at those ignorant, arrogant idiots who believe that they know better than anyone else, and, therefore, do everything in their power to force their beliefs and practices on the entire world.

Enough of this, ... let's move on to some facts of our own. Since Al Gore is the main obstacle to approaching climate change intelligently, he will be my target. If a majority of us become aware of the fraud he is perpetrating, I believe the entire issue will go away.

The information I'm going to share, came mostly from the article, "Thirty-Five Inconvenient Truths: The Errors in Al Gore's Movie," written by Christopher Monckton of Brenchley for the Science & Public Policy Institute (October 2007).

Thirty-five significant errors, and still he's touted as a savior of the world, garnering a Grammy, Oscar, and the Nobel Peace Prize. The rational is that he's done more to bring climate change to the attention of the public than anyone else in the world; and therefore deserves all the awards and accolades he can garner—even if he did it by stretching the truth beyond the bounds of reason; probably even consciously lying to us, for his own personal gain.

Here's a list of the most egregious errors (my opinion) found in Al Gore's movie:

- Sea-level rising by 20 feet, which will cause massive destruction and loss of life, with most of Florida and Manhattan Island under water.

Based on the IPCC's (Intergovernmental Panel on Climate Change) 2007 report, the maximum we can expect sea-level to rise in the next century is 1 foot 5 inches; so Gore's 20 feet is a huge exaggeration. Gore also suggested that there is a mass evacuation taking place on the Pacific islands, due to the current rise of sea level.

This is not true, but environmentalists will do anything to support this claim; even remove a tree from the shoreline so it couldn't be used to show that sea levels are not rising (this happened in Maldivia).

- Increase in CO_2 emissions is causing the global temperature to rise

This is one of Gore's most egregious falsehoods. He shows a graph that demonstrates a direct relationship between global warming and the amount of CO_2 that is in the air. What he doesn't tell his audience is that a close look at the graph seems to say the exact opposite; it's the rise in global temperature that is causing the increase in CO_2. Scientists believe that the rise in CO_2 levels is due to the rise in the temperature of the ocean, which releases more CO_2 as it gets warmer.

- The polar bears are dying.

I've already covered this one in-depth.

- Severe tornadoes are getting more common.

Actually, the exact opposite is true. The rate of severe tornadoes has actually been falling over the past 50 years. However, tornadoes appear to be happening more often. That seems to be result of better reporting and more effective communication, rather than an actual increase in tornadoes.

- The arctic is warming faster than the rest of the world.

Not true. In fact, the arctic has been cooling over the past 60 years; it is actually one degree cooler than it was in 1940. Recently, the media reported that the North-West Passage was open for shipping for the first time since records began (about 30 years ago). However, what they didn't mention was the ships that were icebound in the spring of 2007; or, that it was also open to shipping in 1903 and 1945.

- Mosquitoes and disease are on the increase because of global warming.

Mosquitoes and disease don't need warm temperatures to increase. Some of the world's most dangerous outbreaks of Malaria (mosquito-born disease) have occurred in Siberia in the 20s and 30s. Other diseases that he attributes to global warming are not spread by temperature change, but by rats, chickens, pigs, poor hygiene, etc.

What's New about Climate Change?

- The European Heat Wave of 2003 killed 35,000 people.

Heat waves occur naturally and are not a result of climate change. In addition, there is no mention of the number killed annually because of cold temperatures. In the United States, researchers estimate that 173,000 people are living today because of fewer instances of extreme cold weather. However, we have just experienced a very cold winter and spring in the U.S. ... Where is global warming when we need it?.

- Carbon Dioxide is a pollutant

Actually, it's a food for plants. All plant life thrives in environments that include high levels of CO_2. Actually, when we look at the climate record of the Earth over millions of years, we find that CO_2 levels have been as much as 18 times what they are today.

To add insult to injury, the opening of Gore's film, showing magnificent ice walls, was computer generated for the film, *The Day After Tomorrow*. This footage, stolen by Gore, has no representation in real life, only in the imagination of its creator.

Now, maybe you understand why I am angry. It is criminal for an individual with his position in our society to knowingly perpetrate a fraud on the American people; actually the people of the world. He should be brought before a court of law, and made to pay the maximum penalty for his deception. Will this happen? I doubt it, he seems to be above criticism … well … high priests tend to be above criticism. But, he's in for a surprise if he expects his reputation to escape the criticism of history. Future generations will learn of his misdeeds and the name Gore, will, I suspect, become very appropriate.

Let's take one additional look at this global warming issue; this time, I'll use charts.

Figure 1 shows global temperature from 1988 to 2008. As you can see, there was a spike in 1998, but the temperature quickly dropped and has remained essentially level since then.

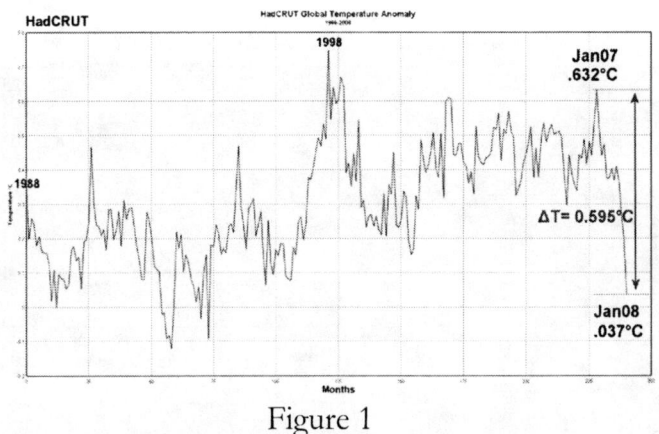

Figure 1

What's New about Climate Change?

Figure 2 shows the earth's temperature from 1880 to 2005. As you can see, the temperature has been steadily rising since then. If you only looked at the graph, you would quickly assume that we are on the path to a much warmer world. This is the chart that environmentalists love to show.

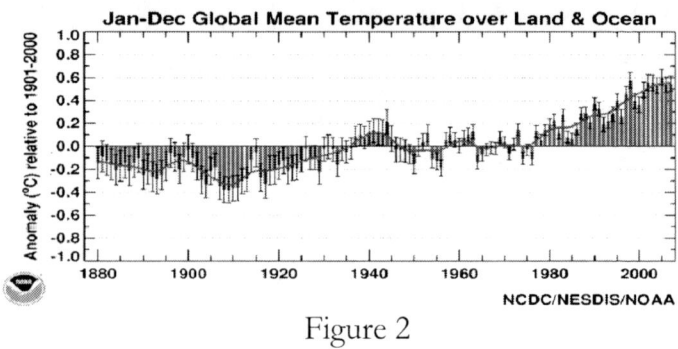

Figure 2

Figure 3 shows the earth's temperature for the past 2000 years. As you can see, the earth is in a continual state of climate change and we are currently coming out of a "little ice age," that was at its worse from 1500 to 1800.

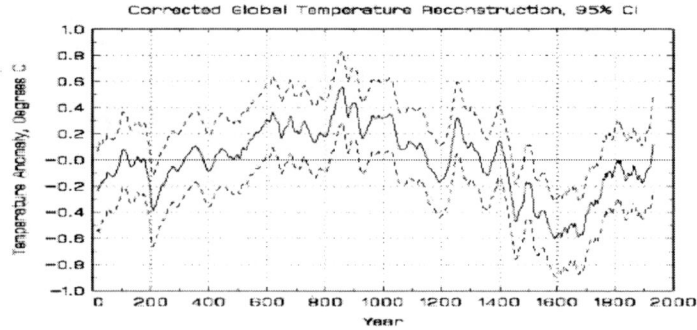

Figure 3

Now, are you still convinced that climate change is a problem, caused by us and our selfish use of natural resources? If you still believe we have to do something drastic to "fix" a natural process far beyond our control, … well, that's your prerogative. …

What's New about Climate Change?

5
A Solution for the Immigration Crisis

The idea that the United States is somehow not living up to its heritage of welcoming immigrants is very wrong; the United States is, in fact, continuing to be the most welcoming nation in the world.

We are a society of people who care about doing what is right, and the right thing to do is to bite the bullet and take care of our current immigration issue. It seems to me that the major difficulty involves those on the two opposite sides of the issue:

- Those opposed to amnesty for "illegal immigrants" (amnesty: the act of the government granting a general pardon for an offense before a trial or conviction, especially to a large group).
- Those who believe that "undocumented immigrants" should have all the rights of citizenship, as well as a guaranteed path to citizenship.

Neither of these positions is reasonable.

We are not going to fine, jail, or deport everyone who is not legally here. The second group won't allow this, business won't allow this, and, in my opinion, the majority of our citizens will not allow this. We, as a society, have a responsibility to the people who have come because we created the situation that brought about this issue.

Neither are we are not going to give immigrants all the rights of citizenship, nor are we going to create a guaranteed path to citizenship for those who are here illegally. I believe the majority of our citizens are concerned about fairness as much as the first group is concerned about following our laws.

We will never solve this problem as long as we insist that either of these two positions be met. And, we will never have peace if we meet the requirements of one group and not the other.

This I know for sure. We must solve the problem. We must secure our borders. And, we must accept responsibility for the situation as it exists. We voted for representatives that created this situation, and now we must deal with it.

An Intelligent Solution to the Immigration Problem

First some definitions:

- Immigrants: People who, while not United States citizens, are living in the United States legally.

- Undocumented immigrants: People who are not United States citizens and are currently living in the United States without legal immigrant status.
- Illegal immigrants: People who enter the United States illegally after the instigation of the program proposed in this paper.

Immigration Facts

The periods of highest immigration were from 1904–1914 and 1989–2006, when the United States welcomed in an average of 1,000,000 immigrants per year. From 1915–1988 the average was approximately 400,000 per year (the years of the depression and World War II were the lowest with immigration falling below 100,000 per year).[1] These figures do not include the 2.8 million undocumented immigrant workers that were legalized by the Immigration Amnesty Act signed by President Ronald Reagan in 1986. Actually, it was the 1968 Immigration Act, signed by President Lyndon Johnson (ending discrimination based on race, place of birth, gender, and country of residence) that started the current "flood" of immigrants which has not abated in spite of the events of September 11th, 2001.[2]

---

[1] From Homeland Security, 2006 Yearbook of Immigration Statistics
[2] From the Center for Immigration Studies

A Solution for the Immigration Crisis

In the two decades around the installation of the Statue of Liberty, the United States welcomed approximately 500,000 immigrants per year. So, as far as total numbers of immigrants, the United States is welcoming many more people today than were welcomed then. However, this is a touch misleading. As a percentage of our population, the numbers of immigrants from 1840 to 1920 were much higher than today. During that time the percent of new immigrants to become U.S. citizens was approximately two and a half times what it is today. However, it is unreasonable to assume that the United States can continue the same rate of immigration decade after decade.

Mexico is the largest source of United States immigrants followed by China, the Philippines, and India. These figures do not include refugees[3]; nor do they include undocumented immigrants. *The United States accepts almost as many immigrants as the rest of the world combined and more than twice the number of refugees as the next nine countries combined.*

---

[3] The Immigration and Nationality Act defines "refugee" in Sec. 101(a)(42) as:

(A) any person who is outside any country of such person's nationality or, in the case of a person having no nationality, is outside any country in which such person last habitually resided, and who is unable or unwilling to return to, and is unable or unwilling to avail himself or herself of the protection of, that country because of persecution or a well-founded fear of persecution on account of race, religion, nationality, membership in a particular social group, or political opinion,
… From U.S. Citizenship and Immigration Services

How many immigrants become naturalized citizens of the United States? The answer is approximately 50 percent of immigrants file for citizenship. This, of course, is much higher than any other country in the world. In fact, many countries do not allow any immigrants to become citizens; this is very true of all Mideast countries, who do not allow even fellow Mideast citizens to become citizens of another Mideast country. In the Mideast, newborns become citizens of the country of their ancestors, not their country of birth.

Objective

This is a proposal that enables the United States to secure the border, and identify and register all undocumented immigrants within 30 days of instigation. In other words, the United States can solve the major issues—the registration and border problem—in one 30-day period. In addition, I have proposed basic solutions to other issues regarding undocumented immigrants. I believe these solutions would be acceptable to the majority of U.S. citizens.

Additionally, while there may be other benefits to following this proposal (90 percent of the cocaine that enters the United States is trafficked through Mexico, Mexico is the United States' largest foreign supplier of

A Solution for the Immigration Crisis

marijuana, and 99 percent of all methamphetamine produced in Mexico is exported to the U.S.[4]), this proposal is focused on the two issues of controlling the border and handling the millions of undocumented immigrants currently residing in the United States.

I believe this proposal offers an opportunity for consensus[5]; a complete solution that a large majority of Americans can support, even though they may not agree with all of the specifics.

First, those of you who believe that any of the following statements are non-negotiable might as well stop reading. I do not believe consensus can be reached on these issues in a reasonable time frame, and, therefore, cannot be part of a solution that can, and should be, implemented immediately.

- The United States should not have limits on Immigration; the United States needs to provide a haven for any and all who wish to come here.
- Undocumented immigrants must be offered a path to citizenship; we owe them this opportunity for numerous reasons.

---

[4] From Foreign Exchange PBS television program, February 22, 2008
[5] Consensus: agreeing to support a decision even though you do not completely agree with it. Consensus is critical because it is impossible to create a solution that will have universal agreement.

- Undocumented immigrants have a right to the same benefits and protections as the citizens of the United States have.
- All undocumented immigrants currently in the United States should be deported and forced to reapply for immigrant status.

This proposal assumes the vast majority of United States citizens believe that our country has a responsibility to those individuals and family units who are currently undocumented immigrants. This is due to the fact that, as a society, the United States has created the environment that has resulted in the current situation. However, the United States does not necessarily owe undocumented immigrants a path to citizenship, or the rights and privileges of citizens of the United States.

The Criteria

The following are essential criteria for an effective solution to the immigration issue:

- The undocumented immigrant problem must be solved in a way that is fair and reasonable. This means the concerns and issues of the undocumented immigrant must be taken into account, along with the public's concern about the necessity of following the laws, and the needs of companies and organizations

A Solution for the Immigration Crisis

for labor to provide the products and services our society needs and/or desires.
- Accusations of racism directed toward fellow U.S. citizens, who are genuinely concerned about the rule of law, must be stopped. It must be accepted that these are honest, caring people who believe that breaking laws is unacceptable regardless of ethnicity or reason.
- It must be acknowledged that being against illegal immigration does not necessarily mean being against immigration. Indeed, the vast majority of U.S. citizens understand clearly the advantages of having a robust immigration process.
- It must be accepted that it is morally correct to challenge a bad law, but that ultimately the citizens of the United States will decide if an existing law is flawed and must be changed. Until such changes, those (individuals, companies, organizations, and local governments) who break the law must accept the consequences of their actions.
- It must be recognized that the U.S. is the most welcoming country in the world, receiving into its boarders the majority of all the immigrants and refugees of the world, and making almost as many new citizens of immigrants and refugees as the rest of the world's countries combined.
- There must be a separation of the undocumented immigrant issue from the national security issue. Directly relating the two is only causing confusion and exacerbating the problem.
- It must be recognized that since the vast majority of undocumented immigrants are entering the United

States by crossing its boarder with Mexico, that border must be the first priority.
- A solution must be developed that enables all undocumented immigrants, whose only crime is being undocumented, to become immigrants within the shortest possible time frame.
- All immigrants must commit to following all laws regardless of their own thoughts or religious beliefs.
- Immigrants who commit misdemeanor or felony crimes will be prosecuted, judged, and sentenced, if convicted, according to the same laws as citizens who commit similar crimes. Once the immigrant's sentence is completed, the immigrant will be deported with no opportunity to return to the U.S. The United States has no moral responsibility to immigrants who harm its citizenry through criminal activities.
- It must be accepted that those already following current guidelines in applying to become immigrants or U.S. citizens have priority over those who become immigrants under this plan.
- It must be accepted that undocumented immigrants who become immigrants under this plan, and have an interest in becoming U.S. citizens, will have to meet all requirements for citizenship.
- It must be accepted that there will be many immigrants who have no intention of becoming citizens.
- It must be agreed that immigrants have the right to distribute their net earnings according to their desires; if this means sending much of it outside the United States, so be it.

A Solution for the Immigration Crisis

The following issues are also part of this plan:

- There is no need to fine undocumented immigrants; there is only a need to document their presence in the United States. In addition, any fees associated with becoming immigrants should be reasonable and no greater than any other applicant for immigrant status would pay.
- All immigrants must have an immigration card that includes the immigrant's name, place of residence, place of employment (if applicable), and nation of citizenry. The immigrant will be responsible for keeping this information up-to-date.
- Immigrants will not have the automatic right to government paid education, social services, and health care. These services will only be provided when reasonable, i.e. K-thru-12 education for immigrant children and reasonable emergency room services.
- Immigrants will be eligible for all employee benefits offered by the employing organization, including health care, paid holidays, and equal pay for equal services.
- Companies, organizations, and individuals currently employing undocumented immigrants will not be penalized as long as their undocumented immigrants are registered under this plan.
- The right to vote will be limited to only those U.S. citizens who are allowed to vote. To assure that this law is followed, voter registration will demand,

in addition to current requirements, proof of U.S. citizenship.

The Process

Once the citizens of the United States have agreed to support the criteria above, getting the immigration problem solved is relatively easy. The critical steps are:

1) Secure the border. The United States must immediately use all of its available resources to make sure that illegal entry is minimized as much as possible. This includes the Border Patrol, National Guard, local law-enforcement agencies, and whatever technology we can put in place.
2) Immediately deport anyone entering illegally; take them to a drop-off point and put them back into Mexico. Remember, initially, this is the high-priority border; this is where our major effort regarding illegal immigration will be expended.
3) Allow a 30-day period for all undocumented immigrants currently residing in the United States to register as immigrants.

Applying for and receiving immigrant status:

1) The United States government will contract with all of the major event-ticketing agencies in the United States, having them provide an "Immigration Ticket"

to every undocumented immigrant, or undocumented immigrant family unit (defined as: husband, wife, and naturally-born or legally-adopted children) with a priority number for signing up with the Unites States government for legal status. Until the actual government meeting, their Immigration Ticket is their "proof" of legal status. The ticket shows their name, residence, place of employment, and contact information. If it is determined that there needs to be a charge for the Immigration Ticket, it will be a reasonable charge.

2) Undocumented immigrants will be given 30 days to obtain their Immigration Ticket. A massive advertising campaign will assure every undocumented immigrant that they will be treated fairly and that there should be no concern for government backlash. Immigrant leaders will also be assured of this so that we have their cooperation in encouraging the undocumented immigrant workers to sign up. Not signing up within the 30-day period, will result in a change of status to illegal immigrant and their ultimate deportation.

3) The government will receive the information from the ticket and will contact each individual or family unit, as to when their meeting will be held with the Department of Immigration; at which time they will receive their official immigrant status.

4) Undocumented immigrants who are currently unemployed will have 90 days from the time of their meeting with the Department of Immigration to find

employment. If employment is not found, they will be deported.

5) The Department of Immigration will hire additional staff so that they are able to process the millions of individuals and family units as quickly as possible. The final immigration card will be their identification and will enable them to open checking accounts, find work, get a driver's license, etc. The Immigration Ticket will serve the same purposes until the immigrant receives their official identification card.

6) The Department of Immigration will notify the media when they have completed the registration of all immigrants. At this time, all Immigration Tickets will be voided.

7) Any undocumented immigrants who have warrants for their arrest (excluding parking or speeding tickets), or are considered criminals in the native country, will not be able to receive immigrant status.

8) Any illegal immigrants caught after the 30-day sign-up period for the Immigration Ticket, or after the Department of Immigration announces that they have registered all undocumented immigrant workers, will be deported to Mexico, or their home country, immediately.

9) Companies, organizations, and individuals employing illegal immigrants after the 30-day sign-up period, will be fined. The fines will be significant, making it extremely undesirable and unprofitable to hire illegal immigrants.

Conclusion

It is understood that, going forward, there will be official requirements for receiving immigrant status; this proposal does not address the specifics of any following rules and regulations. It is important for any rules or regulations developed to take into account the need for immigrant workers; and the government must see as its responsibility the supplying of these workers to those companies, organizations, and individuals in need. Likewise, it will be the government's responsibility to assure United States citizens that their ability to be employed will not be negatively impacted by the immigration worker program.

There will be a system set up in Mexico, and possibly some other countries, where employers can contact government agencies (or private agencies contracted by the government) to recruit needed workers. Workers with assurances of employment in the United States will receive their immigration papers quickly. Any research needed to assure National Security authorities that the new immigrants are not a threat to U.S. citizens or national security will be handled with a high-priority status. It will be recommended that the recruiting agencies vet (get approval for) all potential applicants well in advance. In other words, potential immigrant workers should be identified prior to their being requested by U.S. companies, organizations, or individuals.

Finally, we need to change our laws so that children born in the United States of illegal immigrants would not automatically become U.S. citizens. It is not acceptable that individuals can sneak into the United States to have their babies, guaranteeing that their children will become citizens of the United States. This is the ultimate reward for breaking our laws and cannot be allowed.

A Solution for the Immigration Crisis

6

What Is Intelligent Design?

It's interesting that what gained my interest was Ben Stein's movie, *Expelled: No Intelligence Allowed*. I, like most of you, had always assumed that Intelligent Design was just another name for Creationism. Ben Stein set me straight.

This is one more issue that the media has really screwed up. I blame the media, because they have a responsibility to do the necessary research. Well, this time most of them don't have a clue.

First, let's define some terms that we must have agreement on:

- Creationism: The belief that the world was created by a Living God between 6000 and 10000 years ago. And, that God is still very active in the Universe and our individual lives.
- Darwinism: The belief that life has evolved without a God through a process, Natural Selection, that Darwin postulated in his book, *The Origin of Species*.
- Darwinist: People who believe that Natural Selection is the only process involved in the creation of life and new species. Darwinists are, essentially, atheists.

What is Intelligent Design?

- Atheists: People who do not believe in God.
- Theists: People who believe that there is a God, a conscious force in the Universe.

Theists come in many forms; the spectrum would include: those that believe God started things going, but may not be directly involved at this time; to those that believe that there is something going on that can only be explained by the existence of some sort of conscious force in the Universe; to those that believe that God is still very much involved in everything that goes on in your life; passing judgment on how you are living it, on the decisions you are making, and on whether you will spend the rest of eternity in heaven or hell.

Creationists are the most conservative of the theists; while the agnostic, someone who isn't quite sure what's going on but isn't ready to toss God under the bus, is the most liberal. The beliefs of the vast majority of us fall somewhere on this spectrum.

However, from all appearances, the media see's only two choices, you are either a Darwinist/Atheist or a Creationist. In other words, you either believe that God doesn't exist now and has never existed, or you believe that the world was created between 6000 and 10000 years ago by a Living God that has a personal stake in our lives.

There is no place for those of us who believe that something is going on that can only be explained by a conscious force in the Universe. We may not be able to define that force, but we have faith that it exists.

The media's attitude is extremely narrow-minded and has led to a great amount of confusion, confusion that has led to some wrong-thinking decisions, where people who have a faith (in some sort of higher being or consciousness) find themselves supporting the atheists (regarding the issue of intelligent design), because they cannot align themselves with the Creationists.

Most of us agree that it is naïve to believe that God created the Universe 6000 to 10000 years ago. To believe that, you have to believe that the entire Universe, everything from fossils to the distant stars is a fraud, designed to test the strength of our faith. I don't know what this conscious force in the Universe is, but I do know that it's not a fraud, and wouldn't play these types of games. Because I am a thinking being, I cannot align myself with the Creationists.

But, neither can I except the premise that everything I have seen, everything I know, everything that I have experienced is the result of some cosmic accident or series of accidents.

The result: I'm left out of the discussion, as are the millions that agree with me.

What is Intelligent Design?

First, since you've probably haven't seen it, let me summarize the important points from Ben Stein's movie.

The main message of the film seems to be that in order to be recognized as a creditable scientific researcher, you must accept that the process for creating life, creating new species, and creating new types within species, is essentially a process very similar to Natural Selection. There is not, nor ever was, a universal intelligence involved. If you don't accept this as fact, your personal intelligence is suspect and, therefore, also your credibility.

I'm not sure that this judgment is pervasive throughout the scientific community. I know of many scientists who have a strong faith. To support this contention, Ben had some scientists that stated they had been "black-balled" for seriously suggesting that the possibility of an intelligent source for the creation of life, was worth looking into.

There was no sense that any of these scientists were Creationists, only that they felt that the search for an intelligent source was as important as the search for a non-intelligent process. There is no doubt that we, and I'm including the world scientific community, haven't a clue as to how life first began; or was created (sprung forth) from non-life.

Ben also spent a lot of time showing us where the Nazis had done some of their research, and then aligning the Nazis with the Darwinists. His contention was that a belief in the ultimate power of Natural Selection, leads to a conclusion that, with our current knowledge, we can take control, even speed up the process, and develop a super race.

It's pretty clear that that was what the Nazis were about, and it is well-known that they embraced the concept of Natural Selection; that they were, therefore, Darwinists. This is why they euthanized all who did not meet their requirements, i.e. Jews and the mentally and physically disabled. Since Ben is a Jew, I can easily understand his concern that Darwinism could lead others to make the same decisions—for the good of our species.

It's also very clear that those who support Darwinism would take great offense at this suggestion.

I've made the contention that the media cannot see beyond Darwinism and Creationism, completely ignoring those of us who believe in a conscious force in the Universe. A conscious intelligence that, at the very least, got the ball rolling—created life in the beginning. Therefore, they call anyone seriously suggesting the concept of

What is Intelligent Design?

intelligent design a "creationist crackpot" who shouldn't be listened to.

Let me prove my point; below are a few quotes from the media, reviewing Ben's movie. In reading them, remember that Ben's movie never suggests any of the concepts embraced by the Creationist, only that for life to begin, there probably was an intelligence involved.

Frankly My Dear... Movies with Roger Moore
OrlandoSentinel.com

> How do you re-package that tried and untrue, untested and untestable faith-without-facts warhorse, "Creationism," after its nearly-annual beatdown by an increasingly exasperated scientific community?
>
> After you've tried renaming it "Intelligent Design," ... Maybe Stein will repackage himself as the new face of Creationism.

Scientific American, John Rennie

> Unfortunately, *Expelled* is a movie not quite harmless enough to be ignored. Shrugging off most of the film's attacks—all recycled from previous pro-ID works—would be easy, but its heavy-handed linkage

of modern biology to the Holocaust demands a response for the sake of simple human decency.

Expelled wears its ambitions to be a creationist *Fahrenheit 911* openly. …

Expelling All Reason, Dan Whipple
Committee for Skeptical Inquiry

The film, more than any other creationist/ID effort I've seen, is antiscientific and antirational. In it, Frankowski (the producer) opposes not just evolutionary theory but the scientific superstructure built in the West since Renè Descartes. Using Darwinian evolution as a springboard, he attacks nearly every scientific discipline and the scientific method as leading inevitably to atheism and global evil.

Below I give credit to someone who got it right:

National Review Online, Dave Berg

Dawkins (well-known Darwinist scientist) dismisses the Emmy-winning actor as having "no talent for comedy." He believes during the interview Stein is an "honestly stupid man, sincerely seeking enlightenment from a scientist." A lawyer, a law professor, an

What is Intelligent Design?

economist, and a speechwriter for both Nixon and Ford, Stein hardly seems to fit the description "honestly stupid."

In the end, the film isn't really about intelligent design as much as about a relentless attack on an authentically free inquiry. As Ben Stein points out, "Freedom of inquiry has been greatly compromised, and this is not only anti-American, it's anti-science. It's anti-the whole concept of learning."

The bottom line is, should we not teach in our public schools' science classes that there are two basic thoughts regarding the beginning of life?

- Some sort of accident, a chain of improbable events, came together and life began, fortuitously leading to our existence. (Given an eternity, this is probably possible); or
- An intelligent force, consciousness, created life purposefully. (This begs the question, "Where did "God" come from?)

Darwinists, as shown in Ben Stein's movie, reject the concept of an intelligent force and try to explain everything from a Natural Selection perspective. It seems to me that there is at least as much faith in this belief as there is in the belief that an intelligent consciousness was involved.

The truth is, we just don't know.

I guess it would be okay to state in high school science classes that we don't have a clue as to how life originally began, nor, are we very clear on how new species are created. These are the two controversial issues. Darwin's explanation as to how species adapt to the environment is accepted as scientific fact.

Anyone can come up with a hypothesis to explain how life began and/or how new species are created. For example, I believe that new species are created during catastrophic times, asteroids hitting the earth, super volcanoes creating "nuclear winters," etc. During these times all protections are removed and cosmic rays beam down at full strength, causing massive mutations and the creation of new species. Can I prove this? ... No. ... Is it as good an hypothesis as anyone else has come up with? ... Well, in my opinion, it is.

How about the theory of Intelligent Design? Can we come up with an hypothesis that supports that possibility, and yet would be scientifically feasible?

How about this one: prior to the Big Bang, there was no consciousness, there was only a single point that contained everything we now call the Universe. Then, 14

What is Intelligent Design?

billion years ago this single point began to expand, we aren't quite sure why. At that moment, matter, space-time, and *consciousness* were created. The Universe became aware of itself and marveled in what it was and the potential it had. Soon it began to "play" and the galaxies, stars, and planets were formed. Next the consciousness decided to bring life into the Universe; ultimately, beings that lived and walked and procreated. ... Why not?

So, what do you think, should the hypothesis that an intelligent force was involved in the creation of life be taught in our high school science classes? As I've said, in my opinion, that hypothesis is as viable as saying that life was an accident.

    Maybe we should do neither, maybe just say, "We don't know how this happened—that's what science is all about, finding out how the Universe works. However, if we really want to find out ... we probably shouldn't denigrate any course of action. We really never know where the answer is going to come from.

Here's a story that I think fits well at the end of this chapter. The beginning of the story is fairly well known; I've added my own ending.

A Tall Tale of Turtles

At some time in the distant past, humanity forgot that the Earth hung in space and revolved around the Sun. The Masters of that time met together to develop an allegory to explain what was holding up the Earth. They decided to tell people the Earth rested on the back of a giant turtle.

As the ages went by, people forgot this was an allegory and later Masters taught it as a Truth. Then one day, a student came to one of the Master's with a question, "What, Master, is the turtle standing on?"

The Master thought for a moment and then said, "On the back of another giant turtle."

The student thanked the Master and went on his way. However, a week later, he was back, again asking, "Master, on what is the second giant turtle standing?"

The master again thought for a moment, and then replied, "On the back of another giant turtle."

Again the student went away, and again, he was back a week later. However, this time, before the student could even ask his question, the Master looked into his eyes and with a deep wisdom in his voice, said, "My son, it's turtles all the way down."

As the years passed, more and more students began to question the concept of "turtles all the way down." In fact,

What is Intelligent Design?

many of these questioning students became Masters themselves and found it hard to give this explanation to the students who approached them.

Finally a meeting of Masters was called, and it was decided the turtle explanation would no longer suffice. They would have to come up with another explanation to meet the demands of the students. After days of discussion and argument, they decided their story would be that God, after creating the Earth, had no place to set it; so He hung it in space, in the exact center of the Universe. Therefore, the rest of creation revolved around us.

The Masters knew this was only a story they had made up to satisfy the curiosity of their students. But they also knew if they could satisfy their students' curiosity regarding where the Earth was, they could more quickly move them to the greater questions of what existence was and who they were.

As the years went by, even the Masters forgot they had made up the story; ultimately it, too, became a Truth all were expected to believe.

Then one day a student came to his Master and said, "Master, I have used some new instruments that have come into my possession and made some measurements. These measurements seem to show that the Earth is not the center of the Universe ... what am I to make of this?"

And the Master said, "There is nothing that evil will not do to confuse us. These instruments are the product of an evil one who has given them to you to cast doubt on the Truth, to lead you from the true path. Take these instruments ten miles out to sea and cast them overboard."

The student did as his Master commanded.

What is Intelligent Design?

7
Effective Leadership

Much of the information in this chapter is from my book, *Get Things Done: Ten Secrets of Creating and Leading Exceptional Teams*. This chapter speaks to the crisis in leadership the world is experiencing. There is a pervasive lack of understanding on how to recognize and develop Effective Leaders throughout the organization.

My ideas about Effective Leaders have come from years and years of being an Effective Leader. I like to think of myself as a practitioner, rather than an academic. As a practitioner, I work in the real world, noticing what seems to work, and then developing ideas and concepts about how to do things differently.

I have had remarkable success in getting things done, and in every case, the team performed above expectations and the team members often grew more than they had originally believed was possible.

The foreword of *The Living Company* by Arie de Geus. is written by Peter Senge and in it he talks about the difference between the practitioner and the academic,

> This is a book of practical philosophy. It has been my experience that extraordinary practitioners like

Effective Leadership

> Arie can make unique contributions to management thinking … . Unlike academics who write about what they have thought, practitioners think about what they have lived through. Because the source of their thinking is experience rather than concepts, they show how sometimes the most profound ideas are the simplest.

I have spent my life thinking about, analyzing, what I have lived through. I have spent hours determining why one course of action works while another doesn't. Just being competent at a task is not enough; you must know exactly what you are doing and why it is working. This is the difference between a person who does the right thing intuitively and the one who understands the entire process, how it works and why. This is the definition of mastery.

What took awhile to realize was that difficult things couldn't be done alone. You accomplish the difficult, even the impossible, only through coordinating your efforts with others, through the power of an exceptional team.

There were two times when this lesson was brought home to me. Once when there was so much to do that one person couldn't possibly do it alone, and a second time when there was expertise needed that I didn't have and couldn't get.

The first instance was when I was Director of Training and Development for the Atari Corporation. During my eighteen months with Atari, the company went from 3,500 employees to 18,000; with the majority being hired to work in Sunnyvale, California.

In that short amount of time, we hired a full staff of trainers and support personnel; took over a significant part of the Sunnyvale High School (which had been closed by the school district); and developed, implemented and managed workshops and seminars in a variety of subjects, including: company orientation, supervisory skills, management and leadership, communication skills, sales, manufacturing, and creativity. In addition, we put together partnerships with the Community College District and the University of Santa Clara.

We touched, in one way or another, every Atari Silicon Valley employee, and reached out to manufacturing plants in El Paso, Texas, and San Juan, Puerto Rico. Taken all together, this was an impossible task ... one that couldn't have been accomplished without the power of an exceptional team.

The second instance was when I became a computer games producer for Activision, a video and computer games publisher. This was right at the start of the personal computer industry. When you work in a brand new industry, you have to figure out how to

Effective Leadership

do everything from scratch. We didn't know the best way to develop games. Should we have design teams? Or should we let the game developers (programmers) do whatever they thought was best?

Since programmers founded Activision, we were inclined to give them free rein; and then we did our best to market and sell what they developed. However, this didn't give us the control we felt we needed. Someone had to watch the process, be the link between the programmers and the rest of the company.

Initially, like the high-tech industry, we had project managers in this role. But they didn't have the authority to adequately control the process. And then we realized that while technology was at the heart of what we were doing, we were really in the entertainment business. So we came up with the concept of having producers instead of project managers.

I was hired into Activision to create their Corporate Training Department; and one of the first things my boss, Ken Coleman, the VP of personnel, showed me after I arrived was a description of the ideal video/computer games producer. It was a listing of the characteristics, skills, and knowledge the team felt was essential for success in this role.

A year later, we still hadn't found the right person for the job. One day, on a flight back from Los Angeles,

I was reading over the list of characteristics, when all of a sudden I recognized myself.

When I got back to Activision, I went into Ken's office and said, "I think I've found your first computer games producer."

He looked up from his work and said, "You have?"

"I think so. He seems to have all of the characteristics we've been looking for."

"Who is it? Do I know him?" he asked.

"You know him pretty well; it's me."

That stopped him. He just looked at me for a minute and didn't say a word. Then he said, "I'll talk to you about it tomorrow."

A week later I started my job as a computer games producer ... the first one in the world. (Activision was the first company to use "producers"; Electronic Arts, followed shortly thereafter). And all of my concepts of effective management had to be set aside. I had to find a new way to *get things done.*

Prior to this, I was the one with all the answers. When it came to tasks under my supervision, I knew better than anyone what needed to be done and how to do it. I liked people with high initiative; but I needed them to believe that I was more than the boss, I was also the expert. Now that was no longer possible ... I didn't know a damn thing about computer programming, and the programmers were well aware of that.

All of a sudden the successful completion of every project depended on individuals with skills and knowledge I did not, and would never, possess. Once again, I was forced to acknowledge the importance of the team.

We had a remarkable record; ultimately the teams I led produced some of the most successful computer games of that era (in fact, two of the most successful computer games ever, *Shanghai* and *Computer Solitaire*)[††]. At Activision our success rate was so high that products my teams developed were delivering the vast majority of the company's revenue and profits.

Near the end of my time with Activision, after I had announced my decision to start my own business, Ken asked me what my secret was, "How have you been able to create such a remarkable string of successful products?"

After thinking about it for a minute, I said, "Do you remember a few years ago when you told me that I needed to depend on others more ... that I depended on myself too much?"

[††] Computer Solitaire was produced after I started my own company.

"I do."

"Well, when I began producing computer games, I found out very quickly that I <u>had</u> to depend on others ... this was a job I couldn't do by myself. Then I discovered certain people were more dependable than others; and when I worked with the right people, things got done easier, faster, and better. This resulted in my credo for getting things done:

"Find the right people, show them how good they are, and then get out of their way."

"Is that it?" Ken asked.

"Pretty much ... of course, you have to make sure they're aimed in the right direction and they stay on course ... otherwise, that's it."

Nowadays you hear a lot of talk about leadership; and that's all most of it is, "a lot of talk." Many people who talk about leadership (and team building for that matter) haven't the foggiest idea how to go about it. This includes not only the trainers and consultants, but, also, the professors, business texts, and business schools, even the most prestigious of them.

Ultimately you can only learn to lead by leading; but you can learn a lot by talking to, reading about, and learning from, people who have been Effective Leaders and

Effective Leadership

know why they were effective. This is the critical part ... they must know why what they are doing works. The truth is, many Effective Leaders are unconsciously competent; they're effective, but they don't know specifically what they are doing that makes them effective.

Even worse is the number of ineffective managers who don't know they're ineffective. These individuals don't know they don't know, and yet they're confidently teaching others how to be leaders.

Basically, an Effective Leader is someone who gets things done through successfully coordinating the efforts of an exceptional team; a team that he or she has probably created, and definitely nurtured and led.

Effective Leadership is having the ability to pull together the right people so that the best possible decisions can be made, and the highest level of productivity and quality can be achieved, enabling you to get the right things done.

This doesn't mean you're always going to be right or you will succeed all the time, but it does mean you have a much better chance at it.

Let's take a look at the characteristics of the Effective Leader, collected from years of observing both Effective Leaders and ineffective managers in action.

Defining the Effective Leader

An Effective Leader gets the job done in ways that increase the competence, confidence, and potential of the individual members of the team. They build an exceptional team that can accomplish the seemingly impossible (at least the extremely difficult), all to the long-term benefit of the organization.

Characteristics of the Effective Leader

- Understands clearly and is committed to the goals of the organization
- Highly ethical and trustworthy
- Can recognize, recruit, and motivate the best people
- Able to build and maintain critical business and personal relationships
- Knows how to create an environment where people can excel
- Genuinely curious about what and how others think
- Sensitive to cues concerning potential problems and opportunities
- Lacks arrogance around problem-solving and decision-making
- Not afraid to make tough decisions but only after deep consideration
- Highly intuitive and not afraid to trust those feelings

Effective Leadership

- Extremely creative, not afraid to consider seemingly impossible ideas
- Willing to take risks
- Embraces the unexpected
- Accepts responsibility for one's actions (or inactions)
- Compelled to effective mentoring, eager to pass on skills and philosophies
- Uses centering techniques for relaxation and heightened awareness

Robert K. Greenleaf, a practitioner who was Director of Management Research at AT&T, strikes out into new ground by sharing the idea that true leaders don't control teams, they serve them. His ideas can be found in the book *Servant Leadership, A Journey Into the Nature of Legitimate Power and Greatness.*

Actually, I don't recommend the book to my students. Greenleaf is, at his heart, an elitist, someone who believes that there are those people, from specific races, that are born to be leaders, and that everyone else must follow. What I've done below is "cherry-pick" the most interesting things he had to say. For example:

> The servant-leader *is* servant first…. It begins with the natural feeling that one wants to serve, to serve *first*. Then conscious choice brings one to aspire to lead.

The concept of servant-leadership is at the heart of effective management and will become an essential attitude of the Enlightened Company in this century. Greenleaf describes the ideal servant leader—the characteristics that make up such an individual:

- Wants to serve first.
- An unusual openness to inspiration, creativity.
- An aptitude for discovering the Goal.
- The courage to say, "I will go: come with me!"
- The ability to Listen and Understand.
- Knowing what to say and how to say it.
- Being able to back-off and reorient oneself.
- The facility to accept and empathize.
- Able to know the unknowable; intuition.
- Having foresight ... knowing what's coming before it comes.
- Being aware, able to perceive.
- Unusual ability to persuade others, sometimes one at a time.
- The patience to do one action at a time, knowing that in the end great things will be accomplished.
- Being able to conceptualize, to imagine how things can become.
- Committed to healing, to serving.

In many ways Greenleaf is describing Effective Leaders as I see them; however, what is obvious is that he also

Effective Leadership

believes that these leaders are born, not made, and that it is society's responsibility to identify them and prepare them for their leadership role.

I think where Greenleaf goes wrong is in the belief that one chooses servant-leadership only to serve. This is a naïve view that would ultimately result in the failure of servant-leadership. Leaders with this attitude alone would never be able to compete with those who are out to win at all costs. One chooses servant leadership because it is very effective at getting the right things done right. There is no altruism here, only the desire to get things done.

The truth is that servant-leadership is a superior style of leadership, a style that gets the job done in ways that enhance the team, its members, and the organization as a whole. Because of this, enlightened leaders will choose this style; enlightened leaders want to win in ways that are best for the team, best for the organization, best for the long-term.

There is no doubt that the methods to instill these needed characteristics in potential leaders are in their infancy, but to suggest that it isn't possible is akin to giving up the ghost while you're still trying to kick the rails out of your crib. I'm not saying it is easy; after all we're talking about changing attitudes, changing the way individuals perceive and deal with the world around them.

Another blindness of Greenleaf's is his preoccupation with leaders as members of an elite class. To give you some perspective here, Greenleaf wrote his book in the seventies, with an additional copyright in 1991. This book and the ideas presented here are fairly current. His attitude of elitism becomes apparent as he is bemoaning that certain groups of the "dark-skinned and the deprived and the alienated of the world" are trying to solve their own problems. This is what he says,

> ... *some* of those of today's privileged who will live into the Twenty-first Century will find it interesting *if* they can abandon their present notions of how they can best serve their less-favored neighbor and wait and listen until the less favored find their own enlightenment, then define their needs in their own way, and, finally, state clearly how *they want to be served.*

In this way Greenleaf reflects the condescending attitude of elitism that has been prevalent over the centuries: The attitude that only a few are chosen to lead and that these are the elite among us, those that will show the way; that will save us from ourselves.

Effective Leadership

This is, of course, rubbish. There is no one group that has an exclusive on knowledge, intelligence, or wisdom. You can find these characteristics from the towers of Manhattan to the heart of the Brazilian jungle. And, you will also find ignorance, arrogance, and stupidity in the same places.

Effective Leaders do not follow normal practices when creating their teams. They've learned that depending on résumés, work experience, and education is not the most effective way to find the right people.

Believe it or not, I often don't read résumés and I never trust them. I have found I am often led astray by résumés in both directions; that is, either the person is better than the résumé reflects, or they aren't anywhere near as talented as they claim to be.

The real problem comes when the person is better than their résumé … we'll never find this out if we set the résumé aside and never talk to them. This is how we sometimes overlook the best possible candidate without even knowing it. This doesn't mean résumés aren't valuable; just don't depend on them too heavily. I do a phone interview with everyone who applies.

I've been asked, "How can you afford the time to do phone interviews with every applicant? I thought you were the boss."

"I am the boss, but let me ask you a question, Just how important is it that you find the right people?"

The answer is very simple, *there is nothing more important* than finding the right people. Everything you want to do depends on having a team filled with the right people, doing the right things, in the right way.

When it comes to hiring, I trust my instincts more than anything else. But, I am aware of my "prejudicies," so I have those on the team that read résumés, and we interview until we have found the right person for the job we need getting done.

Motivating Exceptional People

Believe in them.

I didn't say, "Give them a sense of your belief in them." I said, "Believe in them." Really believe in them and their ability to do what they need to do so the job will get done.

This kind of belief comes from the heart, from a place inside of you that believes people are trustworthy, until they show you otherwise. I always trust at the beginning, the vast majority live up to that trust. But … when someone shows that they can't be trusted, they're gone—there's no second chances on issues of integrity.

Effective Leadership

Know they are critical to accomplishing the objectives of the team.

You hired them because you believed they would make a difference. If you didn't believe that, why did you hire them in the first place? Now it's time to know they are critical to the success of the current effort. Exceptional people will sense that you know their value, and they will do everything in their power to live up to that expectation.

Trust them with honest communication.

It continually surprises me when I hear an executive say, "We can't tell them the truth; they won't be able to handle it." What makes any of us believe the capability of handling bad news only exists in the executive suite? What an ignorant, arrogant attitude. Experience has shown time and time again that the unknown is much more difficult to handle than any amount of bad news.
    Of course, the other executive blind spot is thinking we can keep bad news a secret. If you believe this, I can assure you that your people are much more capable of finding out the truth than you want to give them credit for.

Respect their opinions by giving those opinions value and due consideration.

You never know where the answer is going to come from … take that as a given. If you've hired the best available people, then their opinions are important and valuable to the success of the team. Don't shut anyone out ... not if you want to accomplish difficult things.

Provide an environment where they can learn and excel.

Of course, much of the way you provide that environment is by doing the above things. Add to this the opportunity to grow in ways that excite and enthuse them, a chance to be more than they ever dreamed of ... it's a tough act to follow.

Strive to make "their work" part of the job ... so they are doing what they are compelled to do while they are a member of your team.

"Your work" is what you're compelled to do. A job is what you do to earn money so you can survive and, hopefully, find time to do your work. The best example is a musician; his or her work is playing an instrument and/or singing a song, composing music, performing, etc. However, it can be a tough way to make a living; that's why so many of them "keep their day job."
If you can determine what someone's work is, what it is they are compelled to do, and make that part of their job ... now that's motivation!

Finally, treat them fairly.

Notice, I haven't even mentioned salary or benefits as motivators? That's because exceptional people aren't primarily motivated in this way. However, they do expect to be

Effective Leadership

treated fairly in regard to these things. So it's certainly important to be sure you are being fair and that you keep any promises made regarding salary and benefits.

While exceptional people are basically self-motivated, one of the best ways to de-motivate is to treat people unfairly. For instance, you might pay other employees, who are contributing less to the organization, more; or you might ask team members to work unreasonably long hours without adequate justification. Finally, you might not live up to your promises, to the expectations you set in the beginning.

I'm surprised at how motivated and deeply committed exceptional people become, when they are part of an exceptional team that knows and understands its mission.

Here's an example of what I'm talking about:

It's the Right Thing to Do

During my time as chief product officer for Dryken Technologies a series of events resulted in our finally running out of options. I had been telling the CEO for weeks that if we got to the point where we only had two weeks of cash left, I would have to tell my team so they could make arrangements. There was no way I was going to let these committed, loyal people go without some warning.

On Friday of the week I realized we only had two weeks left. I got the entire team, including the VP of research and development, together around a picnic table just a short stroll from our offices.

"I promised you at least two weeks notice when I believed the end was inevitable. Well, that moment has arrived. I can't see how we can continue making payroll beyond two weeks from today."

There were nods around the table. They had been expecting this, and the unusual meeting place had already told them something serious was afoot.

I continued, "I want you to know I consider this to be the finest team I have ever worked with. Your loyalty and commitment has been unparalleled. I'm sorry it has to end this way."

The meeting continued with others talking about the experience and how they would remember it for the rest of their lives. Then Jeff Lomax, our senior engineer, said, "What are we going to do about Diamond Athletics? We need three more weeks to get their deliverable ready."

"I'm not sure what we can do. I'm committed to getting it ready for them, but I won't go back on my promise to all of you that the next two weeks was your time to get ready for the major layoff that I'm sure is coming," I said.

"Well, I'd like to do everything we can, so we can ship the product to them. What do the rest of you think?" Jeff responded.

There was quick agreement around the table that it had to be a priority, and the team spent the rest of the time planning how the project could be finished in two weeks instead of three.

Just as I had predicted, on Saturday night two weeks later, I got a call from the CEO telling me our company was finished, "We're out of money, Brad. I'm afraid yesterday is the last day that any of you will be paid. We're keeping on a skeleton staff to sell the IP (intellectual property) and close things down."

"When did you plan on breaking the news to my people?" I asked.

"I thought we could do it first thing Monday morning," was the reply.

"I'm not sure I can wait that long. I'll give Stephen (our director of engineering) a call, but I suspect we will be calling everyone this weekend.

"Fine with me, if you want to do that," and the conversation ended.

As I predicted, Stephen called every Product Development team member that night; while I called Knoxville and talked with Nancy Grady, our senior scientist and the VP of research and development.

About an hour later, Stephen and I were back on the phone together.

"Everyone is fine ... it's not like it wasn't expected. The only concern is that we aren't quite ready with the deliverable for Diamond Athletics. We need one more day," Stephen said.

"What can we do about it now?" I asked.

"Everyone wants to come in tomorrow and get it ready," Stephen explained.

"Tomorrow's Sunday!" I exclaimed.

"We know that," he replied.

"Do they know they won't be paid for coming in, that they will be working for nothing?"

"They know that, too. What matters to everyone is finishing the project; it's the right thing to do."

Everyone worked without pay for six to eight hours that Sunday; because it was so important to them to know they had fulfilled their commitment to the client, more importantly, their commitment to the team and to themselves.

The Importance of Curiosity

Many of us have taken a course on listening at one time or another in our life. If this is true for you, I would guess you were told learning how to listen is critical to learning how to communicate effectively. In addition, you were probably told about or taught the skill of "active listening" and again told that learning this skill was essential to effective listening. Well, I'm here to tell you that many times it doesn't work that way.

One of the frustrations of the training profession is that many of the techniques taught don't work at home. I've actually had trainers tell the participants, "Let me give you a warning; these communication skills I'll be teaching you today won't work with your spouse."

They were right, but the reason the techniques don't work with a spouse is because they don't work

Effective Leadership

with anyone. You're spouse is the only one with the courage to tell you so.

The reason that learning the skill of "active listening" is a waste of time is because you already know how to listen. I'd bet that you do it whenever you're with someone who is saying something that interests you. Maybe it's an influential person who is sharing something of value, maybe it's a new love interest with whom you are enamored ... you can't get enough, you hang on every word. You see, you already know how to listen ... you just choose not to at times.

All the skills in the world won't help you communicate more effectively, won't help you resolve conflicts and solve problems, if you're not interested in listening. What you have to do is change your attitude toward the problem or person; your tendency to jump to conclusions, to believe you already know the answer to the problem, or you already know what the person means or believes.

This becomes much easier when you allow your natural curiosity free rein. When you do this, listen with honest curiosity, the other person senses your sincerity, and that's half the battle right there.

I've been told by my business associates from the Middle East that the Koran says, "Listen with your heart, not your head." This is the way to gain a true understanding of what is going on, a critical step toward problem solving and conflict resolution.

It Just Doesn't Make Sense

A few years ago, my son Jeff worked for Advanced Micro Devices. At the time of this story he was a master technician in charge of a critical piece of equipment that was not performing satisfactorily. There was a "particle" problem causing the machine's yield (the amount of product delivered that met required standards) to be way down.

As the master technician for this piece of equipment, Jeff had the responsibility for solving this problem; more importantly, he also had the authority needed to make it happen. The first thing he did was call in the engineers from Applied Materials, the manufacturer of the equipment. After a thorough inspection, they discovered residue in one of the four tubes that supplied gas to the chamber.

"This is the answer," they told him. "Everything will work fine once we've replaced the tubes with new ones."

"I'm not so sure ... " Jeff said. "Why is that tube dirty? The residue you think is there shouldn't be in that tube; and if it is, then it should be in all four tubes. Why are the other three tubes clean?" he paused, then added, "I'd like to do a little more research before we jump to any conclusions."

Effective Leadership

"You know, Jeff," one of the engineers said, "This type of problem is not all that unusual. I thought it was important to get this piece of equipment back up and operating. Are you sure you want to delay that happening?"

"The last thing I want to do is tear this machine down and rebuild it, only to have it fail again because we really didn't fix the problem. I need to know what's going on and I won't be happy until I do. I'll get back to you when I understand the issue better."

Jeff thought the company that supplied the gas might have some answers so he called his contact there. "It just doesn't make sense to me, why that particular tube would be dirty, or why the other tubes would be clean," he told them.

"I tend to agree with you, Jeff," Cathy, the supplier's contact, replied. "But to be honest, I don't have the foggiest idea what's happening."

"I can't fix the problem if I don't understand it," Jeff said. "If you can't help me, do you know who can?"

"We've got some scientists in Allentown, Pennsylvania. There's one I like a lot. He's real smart but easy to talk to. How about putting you in contact with him?"

"Great! I'd like to talk to him as soon as possible," Jeff said hopefully.

The next day Jeff was on the phone with Dr. Ronald M. Pearlstein, "I'm with you, Jeff; it just doesn't make sense. If it is a one-tube problem, it's the wrong tube ... otherwise, all four tubes should be involved."

"Do you have any ideas what might be causing the problem?" Jeff asked.

"I'm sure it's not what the engineers from Applied Materials assumed. I've got some theories on what might be going wrong, but I'm going to need more information before I can be sure," Pearlstein added.

Jeff's need to understand the problem was already paying off. It would not have been a good decision to just replace the tubes and move on. But he still didn't have the answers he needed.

"I want to get to the bottom of this," Jeff said. "I'm going to bring in all the suppliers involved with that equipment. I don't want to take a chance on missing anything. If you don't know what's going on ... well, I've got to assume that, at the very least, we've got a pretty interesting problem. I sure would like you at that meeting."

"I can't fly out to California ... Is there a way you can teleconference me in?" Pearlstein asked.

"Sure, that won't be a problem."

A few days later they were all in a conference room: Jeff, the engineers from Applied Materials, and the experts from the companies who supplied the gas and the filters, with Dr. Pearlstein teleconferenced in.

While Jeff's curiosity was paying off, it was in the meeting that it would really shine. In a later chapter we talk about the end of arrogance and elitism ... Jeff is one of the best models of this in action I have ever seen. All he cares about is gaining understanding; he doesn't worry about what others are thinking of him, doesn't have to pretend he has all of the answers. If he doesn't understand something, he won't let it go until he does. He just keeps asking questions, questions driven by his natural need to know ... the fire of curiosity that consumes him.

The result ... once he had everyone together in the same room, they discovered a number of major problems,

Effective Leadership

all of which needed fixing. Once they were taken care of, this specific piece of equipment performed flawlessly. And Jeff got the understanding he was after.

However, the major impact from this exercise was that they determined that Advanced Micro Devices was using the wrong filters on a number of their machines. When the right filters were put in use, the yield on all of these machines increased.

This was a true instance of serendipity. It turned out the residue in the tube was not due to a filtering problem; there was really no need for the filter people to be in the meeting. Without Jeff's intense curiosity, his need to understand all of the issues, this hidden problem would never have been discovered and solved.

Postscript:

Dr. Pearlstein called Jeff a couple of months later, "Jeff, I like the way you think. We're involved in a major research project on gas-line purging, but all of our current data is from prototype testing. We need to do some real-world testing and your Fab would be the perfect place. Will you help us out?"

"I'll do everything I can," Jeff replied

A year later, March, 2001, the paper on Dr. Ron Pearlstein's research, "Evaluating Electronics-Grade Gas-Line Purging Requirements," was published in Solid State Technology, an international magazine for the semiconductor industry. Jeff was a co-author.

You know what? I also like the way Jeff thinks.

We are all born with a natural curiosity. This curious nature is a survival characteristic found in almost all animals. Maybe your natural curiosity has been pushed down by those who have told you things like, "Curiosity killed the cat," or "Don't ask why, just do what I say." But it's still there just waiting to be rekindled; it can still be a power that can make you an extremely effective listener and an excellent problem solver.

To solve any problem the first step is to gain understanding. And, the best tool to use in order to gain understanding, is your natural curiosity. What's interesting to me is that four major challenges for management, for the Effective Leader, are conflict resolution, problem solving, negotiation, and decision making; all are handled in essentially the same way.

The first step in every instance is to gain understanding. And, the easiest path to gaining understanding is the effective use of your natural curiosity.

The second step is to determine alternatives. This step, too, is heavily dependent on your curiosity. It is through honest curiosity, from the heart, that you will uncover the best alternatives.

The final step is decision making, or choosing your best alternative, once the choices are clear. In every instance, your natural curiosity plays a big role.

Effective Leadership

The importance of a healthy curiosity to the Effective Leader cannot be over emphasized.

This is about all I'm going to say about Effective Leadership in this book. However, the next chapter on The Enlightened Company, provides some more insight. For those of you who are interested in exploring this territory at a greater depth, I would recommend my earlier book, *Get Things Done*, or my new book, coming out this fall (2008), *The Art of Leadership*.

8
The Enlightened Company

As the value of the enlightened company becomes more recognized, Effective Leaders will be drawn to organizations that reflect that value. Unenlightened companies will be stuck with ineffective management. If they can't become enlightened and learn to identify the Effective Leaders in their midst, in the long run, these companies are doomed.

In order to become enlightened a company, must creatively handle four significant areas: identifying and developing Effective Leaders; achieving environments where creativity and innovation thrive; streamlining the budgeting process; and, defeating elitism. I discuss identifying Effective Leaders in this chapter, streamlining the budgeting process in chapter 10, and defeating elitism in chapter 11.

Below is my definition of the Enlightened Company.

> A company (or organization) as a living entity that understands its responsibility to its shareholders, team members, customers, communities, society, and the world. An organization that serves all stakeholders because it knows that, in the long run, it will only profit by serving, and, because it serves, it remains strong and healthy even through turbulent times.

Enlightened Companies and Effective Leadership will be an unbeatable combination as we continue into the Twenty-first Century. For any organization to reach its full potential, it must be able to identify its Effective Leaders.

Identifying Effective Leaders

Professor Warren Bennis (leadership guru) sees leadership and management as two, very different things, "Leaders do the right things, managers do things right." He acknowledges that companies need both to succeed, but he focuses on the leadership at the top of the corporate pyramid. It seems that everyone else manages the process, makes sure that their team (group, employees, people, … whatever) is doing things right. In other words, organizations have just a few (maybe only one) leaders and a whole lot of managers.

This assumption is wrong and comes from too much analysis and focus on those who run our Fortune 500 businesses, powerful institutions, and governments, and not enough focus on those who are getting things done in every level of the organization. To even suggest that they don't have to think about the right things to do, shows an ignorance of the real world that is appalling. It is in the body of the organization where we will find our Effective Leaders, in the trenches, accomplishing the impossible; this is where a company succeeds or fails in the long run.

This was brought home to me at Atari, where, as corporate director of training and development, I was able to observe both top management and line management in operation. There was no doubt that, in the main, the line managers were extremely competent, while many of the top executives had no idea what they were doing, although confident that they knew more than anyone else.

The failure of Atari is an excellent example of what happens when arrogant, ignorant people rise to the top of an organization (including the U.S. government)—it's only a matter of time before they crash and burn, usually taking the organization down with them, and sometimes, as was the case with Atari, an entire industry.

Are you aware of following two "rules" of business that directly inpact the ability of an organization to get things done?

- The Peter Principle – *In organizations, people tend to rise to their level of incompetence*, and
- Murphy's Law – *Anything that can go wrong will go wrong.*

What is not known is that these rules apply only to unenlightened companies (organizations, government bodies, etc.) and ineffective managers.

Effective Leaders *never* rise to their level of incompetence. Why not? Because they are not depending on their own capabilities, skills, and knowledge, but rather on the commitment, loyalty, persistence, skills, and knowledge of their team. As they move up in the organization, they continue to surround themselves with the people they need to succeed, to get done what needs getting done.

Ineffective managers rise to their level of incompetence because they ultimately gain responsibility over a level of the organization that they can no longer handle alone. The more intelligent and skilled they are, the higher they will go in the organization before this happens. This is the reason that so many CEOs are both highly intelligent and yet unbelievably ineffective as leaders.

What organizations need, at every level, are Effective Leaders, those leaders who do the right things right. These individuals exist in every organization, but most of the time they go unnoticed.

After reading my speech on this subject my brother, Dennis Fregger, was fascinated by the realization that over the years he had worked under both Effective Leaders and ineffective managers.

Brad Fregger

The real surprise for me," Dennis told me, "was that I hadn't recognized the Effective Leaders I had worked with. I had one boss in particular who was an extremely Effective Leader. This was very obvious on the fire ground where he would instantly hold you accountable for your statements by asking probing questions to your off-hand remarks.

"You learned very quickly to own your statements in a mature, responsible manner and that made you better at what you did. The part that was "unseen" was the day-to-day nuts and bolts of administering to his personnel, which included insulating us from the Administration's in-effective management.

"I've always believed that bosses, supervisors, administrators, etc. are a major cause of stress in the workplace. He insulated us from that. When I did have a problem, I'd talk to him and, in a short time, I'd see the solution and that would be that. Everything went so smoothly when he was my supervisor. It was easy, like breathing. I didn't recognize the power of this style of leadership until I read your speech."

My brother identified a major part of the problem: Effective Leadership is hard to recognize because it seems so natural. As my brother said, "...like breathing."

A personal experience illustrates this beautifully:

The Enlightened Company

I was attending a workshop on how to read out loud from the Bible effectively. We were given a Bible verse to prepare, with our final being the reading of that verse out loud to the instructor and our fellow participants. I knew the instructor, and he knew my abilities as a reader ... he gave me the most difficult Bible verse that he could find, and then smiled a knowing smile as he handed the assignment to me. It took me all day and night to figure out how to read those verses in a way that would both inspire and clearly reflect what the author was trying to convey.

At the end of the workshop I was talking with one of the other attendees and he said, "It wasn't fair, I had such a difficult reading. If only he had given me yours ... it was so much easier." This, of course, was the ultimate compliment ... that a very difficult reading came off so easy and natural.

Effective Leaders work like this. They are as easy and natural, or as hardworking, as they need to be. They come in extra when they have to, but don't spend "after hours" time in the office just to make an impression.

Because it seems easy for Effective Leaders, unenlightened management assumes that their job is easier, or that they are lucky to have such a great team. The fact that it was the Effective Leadership style of management that developed the "great team" and made the job seem so easy, never enters the mind of the unenlightened executive. In this way the Effective Leader's accomplishments are often "invisible" to the more senior executives in the organization, while their "commitment" may even be doubted.

This situation is exacerbated when the Effective Leader is subconsciously competent but has not attained mastery; they don't know what they are doing or why it is working. In this instance, they may even see themselves as being lucky.

Ineffective managers are extremely visible because they are working so hard at getting things done. They're often staying till late at night or getting in early in the morning (or both). They are not necessarily trying to make an impression … they have to spend more time in the office, because it takes them longer to get less done.

Management observes this "working harder" behavior and perceives it as commitment and loyalty to the organization. Ineffective managers also exhibit tons of persistence and the ability to accomplish things "in spite of the incompetence that surrounds them."

These characteristics are seen as highly desirable, especially by those executives who are also ineffective managers and therefore understand what it takes to succeed under conditions where they can only depend on their own hard work and persistence.

The Enlightened Company

So how does it happen that, in unenlightened companies, Effective Leaders don't get promoted and ineffective managers do? After all, they are both getting the job done.

Here is a common scenario that shows exactly how and why this happens:

Jim & Bob

Jim is an Effective Leader, his team runs smoothly, the job always gets done with very little fanfare and not much observable effort. Additionally, Jim insists that his key people attend important meetings with him, and when specific questions are asked, he defers to the appropriate expert on his team.

When Jim goes on vacation the company hardly realizes he is gone. When or if a question arises, his peers and his boss, as well as others in the company, have no problem checking with the person Jim left in charge, or the person responsible for the area in question.

Bob is an ineffective manager. His team often seems like an albatross around his neck. The job always gets done, but not without a series of potential disasters that would have been catastrophic if Bob had not "saved the day." Bob does not like to have his people in attendance at important meetings, and, if they must attend, he still does all the talking. Even when members of his team are asked direct questions, he interrupts and adds his own "wisdom." Often at these sessions it is obvious that his team members are not capable of understanding some of the more complex issues for which Bob is responsible.

After the meeting is over the executives agree that without Bob running that group they would be in bad shape, that it's good to have someone of his capabilities around when the going gets tough.

When Bob goes on vacation all Hell breaks loose. Nobody can answer the more complex questions that come up ... which means that the boss is often forced to call Bob and get the answers directly. There have even been times when Bob was asked to come back from vacation early.

After Bob and Jim have been on the job for a couple of years, a management position opened that would be a very nice promotion for either of them. The position opened because a senior manager had reached his level of incompetence and the job just wasn't getting done. The group was perceived as being very weak, unmotivated, with very low morale. The question was, "Who should be promoted to whip them into shape?"

From the perspective of senior management, Jim has not had experience dealing with complex issues, or people who need to be tightly controlled—if the job is going to get done.

While Bob, on the other hand, has had lots of experience dealing with this type of a situation.

There is no doubt in the executives' minds that Jim would be "eaten alive" by this group of malcontents, while Bob would show them very quickly, "who was the boss."

The Enlightened Company

From the executives perspective there is really no choice—Bob is the man for the job.

This is how ineffective managers get promoted until they reach their level of incompetence, while Effective Leaders get left in jobs well below their ultimate capabilities, destined to less financial and career success—to the detriment of both the Effective Leader and the company. This is the hallmark of the unenlightened organization.

In my seminars, where this scenario is treated as a case study, many participants choose Bob, because they see him as the least risky choice. They don't recognize that Jim is an Effective Leader, and that he is demonstrating the most important characteristics of the Effective Leader. Instead, they focused on the group and decided that they needed someone to "whip them into shape."

Bob seems to be the one most likely to do that. In addition, it seems obvious to them that Bob is the committed one, the one willing to come in early, stay late, come back from vacation, etc. Jim, to the management group, seems to be a competent manager, who has probably reached his highest level of competency, and is lucky to have ended up with such a great team.

Brad Fregger

Consequences of Incompetence

This story shows how companies hurt themselves when they fail to recognize the leaders in their midst. This also happened to my son Jeff, a few years prior to the previous story.

Jeff began work at AMD shortly after finishing his obligation to the Navy. With his experience as an airplane mechanic, he was given the opportunity to work as a mechanic at AMD's main, Sunnyvale, California plant.

Within a relatively short period of time Jeff became one of AMD'S top mechanics, one whom even the engineers would come to when they had a tough problem to solve. His boss was terrific, a natural leader that trusted Jeff to do the right thing, at the right time, in the right way. I'm sure that none of the executives had any idea how much money Jeff's boss was saving them; how much Jeff and the other team members were accomplishing.

Then AMD announced their very first layoff. The CEO told the press, and the employees, that nobody of value would be laid off. ... Jeff's boss, a natural leader, whose team was getting a tremendous amount of work done, was laid off. Jeff was devastated.

But that wasn't the end of it. Jeff's new boss was a power-and-control manager who needed to know everything that was going on at all times. He was the exact opposite of the previous manager.

As a result, productivity was half of what it was under the old manager, and the morale of the team was at the lowest level ever.

AMD's failure to recognize Effective Leadership resulted in a dramatic decrease in both team morale and productivity, at the worst possible time. The CEO knew he needed the best people; he just didn't know how to identify who the best people were.

The Pros Know Who the Best Are

In the reading workshop example above, the other participant wasn't able to recognize that it was my skill that made the reading appear to be so easy. But, the leader had recognized this and had been quick to compliment me. In the same way, a symphony conductor can easily tell the difference in skill between the violinist sitting in the first chair and the one sitting in the fifth chair. Likewise, a professional coach knows immediately who the best players are. In most professions, the pros know who the best are, who are the most capable of getting the job done.

It continues to amaze me that, in business, government, organizations of all kinds, so many senior managers are not able to recognize the real pros, the Effective Leaders, that they have working for them.

Brad Fregger

I'm Lucky That Way

This incident took place around 1970. Joe Kipper was the Sears store manager in Mountain View, California when I first met him; I was managing the menswear store next door. We hit it off right away and became fast friends; more than that, he became one of my most important mentors.

Joe had been a store manager for so long that he was a member of the Million Dollar Club, a select group of Sears store managers whose stock was worth well over a million dollars (a lot of money in 1970). Joe was so good, that the last store he managed before his retirement, the Arden Fair store in Sacramento, California, was the highest profit store in the entire Sears chain for two years in a row.

The Sears CEO was so impressed that Joe's store had won the award two years in a row that he flew to Sacramento to congratulate Joe and his staff and present the award in person.

After the award ceremony, Joe rode back to the airport with the CEO. They were discussing the store and what a fine example it was, when he said to Joe, "You know, that's an impressive team you've got there."

"Thank you," Joe responded.

"You know, Joe ..." he said with a conspiratorial tone, "with a team like that, *anyone* could have the best profit store in the chain."

After a moment's thought, Joe said, "Yep, I've always been lucky that way."

By observing these kinds of situations over the years, I have realized that the Effective Leader will never be recognized until our companies became more enlightened. The Twenty-firsts Century will not be as forgiving of unenlightened companies as the Twentieth was; with the rate of change we are experiencing, it is essential that companies recognize the Effective Leaders in their organizations and use them to their maximum advantage, by promoting and rewarding them with the opportunities they deserve.

Learning to identify their Effective Leaders is a company's only chance to survive. We can no longer waste our leadership; it's time to promote the Effective Leaders and retrain, or lay off, the power-and-control managers. In this new age, the winners will be enlightened organizations run by Effective Leaders. This means that today's organizations will change, become enlightened, or die; that managers will change, become Effective Leaders, or be replaced.

9
Achieving Innovative Environments

The ability to be creative and to achieve environments where creativity and innovation thrive is a critical characteristic of the Effective Leader. This is especially true as the rate of change accelerates and we run into situations that we haven't faced before; we find that the old ways of doing things aren't as reliable, might not even work at all.

O ver the years I have often been asked how creativity is enhanced, how people can become more creative, how leaders can achieve environments where creativity and innovation thrive.

For many the whole concept of creativity is a black hole, a mystery understood by the "priests" but not meant to be understood by the "masses." Additionally, many have misconceptions about what creativity is, what makes a person creative. Some believe that creative energy is a talent that you either are born with or never possess. It is seen as a characteristic that can lead to greatness or frustration, maybe even madness. People who are creative seem to have, like God, the ability to create something out of nothing, to work from the blank canvas or block of marble.

Achieving Innovative Environments

So ... let's set the record straight. First, we are all naturally creative; this is a basic survival characteristic that is as much a part of who we are as humans as is the opposable thumb. It isn't just a few of us that have been blessed in this way. This bears repeating ... *we are all creative*, we all have the creative energy needed to solve the problems we face, to bring new and exciting ideas and experiences into our lives.

There is no doubt that our environment can limit, even eliminate, this natural tendency. What's critical is our personal ability to filter out those messages that suggest that we aren't creative, the personal determination to meet challenges and overcome obstacles, to prove to ourselves that this new thing we want to try is worth the effort. All of this plays a role in how creative we *believe* we are. Regardless, we are all born with the capacity to create.

The good news is that no matter how you feel about your ability to create today, you can rekindle the creativity that is inherent in yourself; rekindle the creativity of your team members—achieve the innovative environment that is necessary to doing the right things, right. It is my hope that reading this chapter will remind you of the truth of your own creative nature. Let's talk about how we might rekindle that creative spark.

My first experience with the concept of creativity enhancement was when I was director of training for the Atari Corporation in the early '80s. Atari was the textbook unenlightened company by the time I arrived. Nolan Bushnell had sold Atari to Warner Communications (the precursor to Time Warner); and, in their infinite wisdom, Warner had brought in rag merchant Ray Kasar to run the company, believing that Atari was dealing with a commodity and not a creative endeavor.

This lack of vision is especially hard to understand when you realize that Warner Communications owned among others, Warner Brothers (movies) and Warner Records—businesses dependent on the creative energy of talented individuals.

Ray didn't believe that the talent needed to create a new video game was any more critical than the talent needed to create the design for a new rug. (He might have been right in this belief, but that only means that he didn't give credit where it was due when he was running Burlington Mills either.) For this reason, he refused to allow the video game developers any credit for their creations. All video games were created by Atari; the individual who actually developed the game was a piece worker.

The result ... some of the best, most creative programmers left to start their own company, Activision,

Achieving Innovative Environments

at a time when Atari needed every programmer they could use to create the games the public was clamoring for.

    The situation was critical, Atari needed games, and many of the software engineers assigned to develop them lacked the creativity needed to accomplish the task. As the director of training and development, it became my responsibility to find a way to enhance their creativity.

At about the same time I had become interested in a new counseling technology, Neuro-Linguistic Programming (NLP). I had originally been interested in NLP because I thought that the concepts presented could be valuable within the training and development profession. However, I was quickly drawn to the original process that had been used to develop the technology. Richard Bandler became curious about why some counselors helped people after very few visits (five or less), while others had little impact after many visits (weekly for a year or two). It seemed obvious that the effective counselors were doing something different, something that was having significant impact on the patient ... what was it?

    To find the answer, Bandler interviewed both effective and ineffective counselors. This turned out to be a frustrating experience, because they both described their methods in the same way. Worse than that, the descriptions were nebulous, lacking the specifics needed to determine exactly what was making the difference. It seemed obvious to Bandler that neither group knew what they were doing; the ineffective counselors were unconsciously incompetent, while the effective counselors were subconsciously competent.

Let's take a side trip. (Don't worry, all of this leads to the enhancement of one's creative potential.) While it's relatively easy to accept that someone could be unconsciously incompetent and not know what they're doing, how could someone be subconsciously competent and not know, at least at some level, why and how they are being effective? This situation results when one moves from unconsciously incompetent directly to subconsciously competent without going through the learning cycle.

The steps of the learning cycle are:

- unconsciously incompetent,
- consciously incompetent,
- consciously competent,
- subconsciously competent.

Here's an example: go into a first-grade class and say, "I need someone to drive my car home. Could any of you help me out?" You'll probably get some hands, usually

Achieving Innovative Environments

boys, from those in the class who don't know that they don't know how to drive a car, they're *unconsciously incompetent*. The ones who don't raise their hands know that they don't know how to drive a car, they're *consciously incompetent*. A few years later, in driver's education, you'll find a lot of people who are paying attention to everything around them in a very conscious way, they're *consciously competent*.

Finally, you have those of us who drive into the driveway after an hour's drive from work and can't remember anything about the drive home, almost like the car was on auto-pilot, we're *subconsciously competent*.

The problem is that when you go directly from unconsciously incompetent to subconsciously competent, you don't have knowledge of what it is you're doing. This causes two significant problems:

- If the situation changes, it's difficult to know what to do to make needed adjustments and,
- You can't pass on your methods, the specific ways that you are accomplishing things, to others—you don't have the knowledge.

So, it's very possible to be subconsciously competent and not know, exactly, what you're doing and why it is working. This is where mastery comes in, when you are subconsciously competent in a skill, and know exactly what you are doing and why it is working, you have attained mastery over that skill.

This was the situation Bandler found himself in. His effective counselors had no knowledge of what it was they were doing that made them effective. What to do?

He decided to observe both groups as they worked to see if he could determine subtle differences in their approach. After analyzing hundreds of hours of video tape, Bandler was able to identify a number of specific techniques being used only by the effective counselors. He had no idea as to why these techniques would make a difference, but he decided to teach them to ineffective counselors to see if they became effective … they did, and NLP was born.

In my opinion, creativity is another ability where people go from unconsciously incompetent directly to subconsciously competent. This is true for most intuitive skills, skills we learn as an infant, like walking and talking. However, in this instance I had to teach others this skill and I was having trouble identifying what it was, specifically, they needed to learn.

Most books I read on the subject of creativity were either nebulous about the process, or cutesy, providing "little exercises" to help us learn how to "think

Achieving Innovative Environments

outside of the box" or "express the freedom to color outside of the lines." In my experience, these types of exercises had no long-term value, didn't enhance the individual's basic creativity. And, in fact, often "convinced" some of the participants that they were not creative.

I went to Bandler and told him that I was interested in his process; that I wanted to use it, or something similar, to determine those behaviors, attitudes, whatever, that made someone creative. He introduced me to Bob Dilts and the process was begun.

While it wasn't possible to videotape effective game designers being creative, we were able to develop an interviewing process that enabled us to determine how the effective designers were operating differently from the ineffective ones. From these interview results we developed a workshop to enhance creativity that proved to be even more successful than I hoped. One of the first things we learned was what creativity isn't:

- It isn't beginning from scratch, starting from a blank sheet.

None of the really creative game developers ever started from scratch. They always had inspiration from some place, or something that was eating away at them. It might be an image that suggested a story that suggested a game, or an experience they wanted to represent, or a feeling they wanted to convey. It didn't matter, just as long as it led to a game idea.

Some of the ineffective developers had this vision of creating something brand new that no one had ever seen before. They thought they were "cheating" if they got an idea from something or someone else; they weren't creating, they were copying. This belief turned the creative process into something beyond their (or anyone else's) capability.

Isaac Asimov in his book, *Asimov's Chronology of Science & Discovery*, connects the links between all the significant discoveries of science throughout recorded history, and we, the reader, clearly see that nobody discovers anything without building on what has come before, without getting inspiration from another source.

- It isn't developing the skill of "thinking out of the box," or any other tricks.

Another thing the effective game developers didn't do was think about the creative process, think about what they could do to enhance their creativity. They never said, "Now, I can't be limited in my thinking. I've got to think out of the box on this one." To them the process was a natural one, a way they thought about the problem;

Achieving Innovative Environments

not a way they thought about, the way they thought about the problem.

It seemed to me that in the long run these "tricks," these ways of trying to force creativity, get in the way of being creative, because they don't exercise the right muscles, they don't support the creative process itself.

If this is what creativity isn't ... what is it?

During the interviews we identified five of the critical characteristics of creativity:

- It is innovation.

Using creative processes to produce a product, process, or service that is of value to the human condition; everything from developing a new product to sell, to coming up with a new process that results in orders being filled twice as fast.

- It is being sensitive to unexpected developments.

This is a type of "cue sensitivity," a skill that can be developed that helps identify changes in the situation that could have a significant impact on the outcome if taken advantage of.

Winston Churchill said, "Men stumble over the truth from time to time, but most pick themselves up and hurry off as if nothing had happened." Developing the skill of "cue sensitivity" helps assure that you won't be one of the ones who hurry off, unaware of what has happened.

- It is being curious about, "What would happen if..."

You notice that I didn't say, "You think about what would happen if..." The operant phrase is "being curious about." Curiosity is central to creativity and a characteristic that is essential to living effectively with others. The good news is that this is also a characteristic that is part of the human condition, part of who we all are.

Sure, it's possible that your curious nature has been stepped on a few times in your life. You may even have disciplined yourself to keep your curiosity under control; after all, "curiosity killed the cat." But it's still there below the surface, just waiting for you to let it loose again.

What enlivened my own curiosity was committing myself to wonder, by allowing myself to be awed (or odd...whatever), to revel in the fascination of a line of ants stealing honey from the counter top, to be amazed as I watch a 747 take-off, or a fax being sent into the

Achieving Innovative Environments

ether, only to be immediately received on, the other side of the world. As the punch line from one of my favorite jokes says, "How do it know?"

One day I realized that I had a choice, I could choose to be a cynic, or I could choose to be an innocent. I chose innocence. The world is so full of wonder, I didn't want to miss any of it.

Doug Henning, the world-famous magician who has since passed away, said it this way, "When we're little kids, we're filled with wonder for the world—it's fascinating and miraculous. A lot of people lose that ... Magic renews that wonder."

- It is being disciplined.

Creativity without discipline is chaos. This was an overriding characteristic of effective games developers; they were disciplined. Ineffective developers tended to complain about discipline, "I need freedom to create. I can't have any limits ... I don't work well when people try to manage the process."

In my experience the exact opposite is true; creativity is at its best when it is disciplined, when there is reasonable structure, limits, and/or deadlines. Most people recognize the truth of this ... why else do we wait until the night before to write that paper? Then there is

Japanese Haiku, an extremely disciplined form of poetry that is highly creative (my favorite Haiku, "The heat of the day is best measured by the length of my sleeping cat").

In addition to these characteristics, we also identified the successful strategies used by the competent games developers. Central to the acceptance of these strategies was a belief in the capability of the subconscious mind.

The successful strategies are:

- Take time away, do unrelated tasks.

Very often inspiration came after a problem was set aside consciously, which enabled the subconscious to work the problem without the interference of the conscious mind.

This is critical and speaks directly to a faith in the capability of the subconscious mind. I use this technique when I see an engineer bogged down on a problem that just won't be solved. I give him something else to do, something to take his mind off of what he has been thinking of without relief for hours and hours.

Often, a few hours or a day or two later, I'd see the engineer with a big smile on his face, "What are you so happy about?"

Achieving Innovative Environments

"Remember that problem I couldn't figure out ... the answer just came to me while I was working on that other issue for you."

At times like this, the inspiration seemed to come from a small still voice from within. Effective games developers trusted this voice, trusted the inspiration, and celebrated that moment of discovery.

This is so difficult for ineffective managers to deal with. They insist that their employees "stay focused," not realizing that they are part of the problem, that it is this insistence on staying focused that is slowing things down, making it almost impossible to solve the difficult problems that we run into during the development of new products.

- Seize the moments of inspiration.

Creative people know their moments of inspiration and seize on the ideas that come at those times.

I often have my moments in the early morning hours when I am making the transition from sleeping to waking. Sometimes it feels like I had a dream and that was where the inspiration came from. Other times, it's like an idea flashed in my mind and woke me.

David Crane, the developer of Pitfall (among many others) for Activision, told of getting inspiration in the shower. Again, it doesn't matter, we're probably all somewhat different in this regard. When do you often find inspiration, ideas, coming to you?

- Embrace surprises, coincidence, synchronicity, serendipity.

A book could be written (if it hasn't already) about the multitude of major discoveries that have been accidents, surprises, the result of coincidences. Learning to accept and embrace the unexpected is a critical strategy for enhancing one's creativity. It is a shame how many of us become so focused on our plans that we miss the unplanned opportunities that if acted upon, would make the realization of our ultimate goals so much easier.

As a summary, let's take a look at some characteristics creative people have in common:

- They understand what creativity is.
- They believe they are creative.
- They know what works for them.
- They seize their creative moments.
- They are intensely curious.
- They are disciplined.

Achieving Innovative Environments

- They know the secret of "shifting focus" when they hit a stumbling block.

One of the other things that was extremely important to me was to identify what Effective Leaders could do to create an environment that would encourage and enhance the natural creativity of their people. We came up with four specific things, all of which made a significant difference in how creative the environment was.

- Celebrate the moments of creativity experienced by both individuals and the entire team.
- Encourage those working on significant problems to take a break, think about something else, tackle another, easier, issue, anything to get their mind off the current problem.
- Ask enabling questions, "Why does it work that way?" "What if we tried this?" "What are we trying to accomplish?"

These questions should come from your own curiosity, your heart, your own need to discover what's going wrong. Not from a list of "Questions to Ask When Your Engineers are Stuck."

- Discover your team's passion and encourage it. Do this for both the individuals and the team as a whole.

There is one final thing that has a significant impact on our ability to develop creative solutions to major, paradigm-shifting, potential problems. This is our reticence to face our greatest fears, the things that can happen that can destroy all of our plans. Most of the time we want to ignore these larger issues.

A few years ago it became apparent to me that Silicon Graphics was in great danger of losing their graphics technology lead in the computer entertainment industry. A significant increase in power on the PC side, coupled with superior PC tools for this industry, would threaten Silicon Graphics' claim of graphics superiority.

I had a discussion with an executive in the Games Development Division, but was literally thrown out of her office when I suggested that their dominance in the games industry was in danger of being taken over by the PC. They were not able to face this possibility ... within 9 months their dominance came to an end. This impacted their bottom line dramatically.

As Effective Leaders, we must encourage ourselves, our team, our organization to face our greatest fears. The unvoiced fear is a block to creativity; the faced fear, a stimulus.

Achieving Innovative Environments

10

Streamlining the Budgeting Process

One of the critical characteristics that helps define Enlightened Companies is that they are fiscally conservative at their core, as an essential part of their culture.

Fiscal responsibility as seen by the Enlightened Company, is radically different from the way it is viewed currently in the business world. In fact, the current methods of controlling fiscal activity are actually counterproductive.

Any method designed to control fiscal activity must have as its ultimate goal increased profitability and therefore increased return on investment (ROI). This is not true of the budgeting process now in use by the vast majority of businesses in the United States, and, because of recent changes, around the world.

Whenever I think about the budgeting process, including quarterly reporting, it reminds of a story that I'm sure you heard before.

The Emperor's New Clothes

Long ago, in a land far away there lived an Emperor who was very arrogant. He believed that nobody could tell him

Streamlining the Budgeting Process

anything he didn't already know. At the same time he was very ignorant of many things, and because of this combination of arrogance and ignorance he could be easily fooled.

One day two salesmen rode into the land. These salesmen had heard of the Emperor's arrogant ignorance and knew that they could sell him anything they wanted to, as long as they played to his beliefs about himself.

They passed themselves off as tailors from a land that was even further away than the land of the Emperor. They let it be known that they made clothing only of the finest magical cloth in the world, a cloth that would be invisible to all but the powerful and the wise.

Of course, the Emperor had to have a suit of such wonderful cloth and, of course, he could see it in all its spectacular glory. On the day that the suit was finished the Emperor decided that it was too wonderful to hide, that it deserved to be featured in a wonderful parade, so all of his subjects could see just how magnificent their Emperor was.

As the parade wandered around the town it finally came to a street where a young boy was standing with his mother. Like everyone else, the mother was applauding the Emperor as he walked by in all his glory.

The young boy tugged on his mother's skirt trying to get her attention, but she kept saying, "Not now. Not now. I'm watching the Emperor."

The boy, in his frustration, finally hollered out, "But Mommy! He doesn't have any clothes on!" All the applauding stopped ... there was only silence ... and then, one person started laughing, then another, and another, until everyone in the whole land was laughing.

Let's start by looking at how the current way we do budgets will impact a company's ability to adjust to the coming changes.

Arie de Geus, in his book, *The Living Company*, states fairly categorically that the budgeting process does not work. He does this as an introduction to a section where he talks about the necessity of preparing for the future, and how our current methods are failing us,

> Each year, after countless meetings and reports, and an enormous amount of thought and effort, the board finally reaches its approval. ... Each year, the Unified Planning Machinery delivers its estimates of future activity, and each year, the group as a whole bases its investment decisions on those estimates. There's only one problem. Whenever times are turbulent, and anticipating the future is most critical, the Unified Planning Machinery is wrong—Dead wrong.

Since the rate of change tends to create turbulent times, the current method for planning/budgeting will be in even greater difficulty. Arie doesn't even attempt to come up with an alternative. I can understand this. Our current method of budgeting is so entrenched that it

Streamlining the Budgeting Process

seems impossible that we would ever be able to replace it with something that works better, something that will work in the Twenty-first Century.

I've experienced personally how the current methods for budgeting and planning get you into trouble when you're in an industry that is rapidly changing.

I was the director of training and development for Atari during a time of phenomenal growth (from 3,000 employees to 18,000 in 18 months).

I had the reputation of being a maverick, so when the VP of personnel told us (his team) that we would have to have new budgets ready in two weeks, he knew that I would be against wasting our time and energy in this way. *There's just too much to do ... we must maintain focus.*

Before I had a chance to say that out load, he looked at me and said, "Brad, why don't you see me after the meeting."

I agreed.

After the meeting, when he and I were alone in his office, he said, "I know you're against the process, but I have to submit complete budgets to management in two weeks. I need your support."

"You know I don't have the time to do this properly. In addition, anything we put together will be outdated before you get to your meeting. This is ludicrous!"

He thought for a moment and then said, "How about this? I send Judy (his assistant) over to your office tomorrow to make some notes; one hour maximum, probably less ... If you'll cooperate. Then she brings the notes back here and we prepare your budget for you."

Seemed like a reasonable solution for me, "That'll work. Thanks."

Judy was at my office first thing the next morning, asked me a few questions, and we finished up in half an hour. I couldn't help but think, *This is the way everyone should prepare budgets.*

Two weeks later we (the HR team) were sitting around a conference table at 7 p.m. It had become common knowledge that his assistant had prepared my budget. The rest of the team knew that I thought it was a terrible waste of our time, time we couldn't afford when things were changing as rapidly as they were.

Our boss opened the meeting by thanking everyone for getting their budgets in on time, "I know how hard you've worked and I want you to know how much I appreciate that. However, ..." he paused for a moment, looked down at the budgets in front of him, and then looked over at me, then he continued. "However," he took a wastebasket from the floor, "this is the end result," and tossed all the budgets into it.

There was silence and then the VP continued, "Things have changed, we have all new numbers and we must have completed budgets by the end of the evening ... let's get to work."

Organizations are going to have to figure out how to plan in more effective ways; ways that don't steal the productive hours of their management people, or limit

Streamlining the Budgeting Process

the flexibility needed to deal with the ever-increasing surprises that will come their way. The current budgeting process is an albatross around the neck of management, an anchor strapped to the back of our leaders, one they can't afford to carry around, not if they are expected to get the job done in the Twenty-first Century.

The vision I have here is of the Strong Man Olympics. One of the events involves running a race against the clock while carrying a massive stone that weighs hundreds of pounds.

Are you getting the picture? The strain on the participants' faces as they race down the track, trying to beat the clock, while struggling to carry a stone that two normal men couldn't even lift.

That's the picture I see when I think of managers trying to deal with the current budgeting process while running to keep up with a rapidly changing environment.

Why do we create the massive budgets? Mostly because we are addicted to quarterly reporting, and up-to-date budgets are believed to be central to the reporting process. So, why do we have quarterly reporting?

- So the organization will have a plan for the future?

Nope ... quarterly reporting actually focuses companies on short-term goals. Future plans tend to go by the wayside.

- So top management will believe that they have control over the future?

Probably yes, top management tends to be very "hands-on," and up-to-date budgets tend to give them the feeling of control.

- So shareholders will believe that top management has control over the organization?

Very true, shareholders expect top management to know everything that's going on in the corporation, at all times. This is a bit unrealistic, but ... shareholders are often unrealistic.

- So that team members will be committed to the financial goals resulting from the plan for the future?

Maybe ... sometimes ... however, in my experience team members are much more committed to the ultimate goals of the organization and are frustrated by managements' interference in the process, demanding useless reports and tasks that have nothing to do with their goals and objectives.

Streamlining the Budgeting Process

- So that spending will not get out of control?

A culture that rewards conservative spending, without hampering the purchase of those items needed to get the job done, does this much better than the typical budgeting process.

- So the organization will have a plan to fall back on when things go astray?

I don't think so, as we've said above, it's during turbulent times that the current system lets us down. We might want to believe that reports and budgets will prepare us for the future, but they're really about the past—a picture of the past, that we trust reflects the future.

I can easily imagine a conversation with the top executive of a major corporation. It's lunch time at a local restaurant, we've met as friends. He wanted to relax and get a couple of things off of his chest.

"Hi Brad, glad you could make it."

"Me, too, Jim. How have things been going?"

"Not like I expected. My VP of product development just told me that our newest product is going to be a couple of months late."

"That's too bad. What happened?"

"It's based on the latest Microsoft Windows release and he said it's going to be later than expected. Six months ago he told me it would be ready by now. Everything, our financial plans, sales projections, client meetings, everything depended on that prediction. It's going to be a tough meeting with the investors. Boy, I wish we could learn to do that better. You'd think someone whose been around as long as he has could figure it out."

"I wouldn't blame him, it sounds like he was blindsided by Microsoft. I'm sure he's not the only one."

"Well, ... actually ... to be fair, he told me that Microsoft was unreliable and that we should plan on that. The problem was, the board was insisting on a launch date; I had to give them one."

"Regardless, there's not much you can do about it now. How are things on the personal front?"

"Judy's fine ... well, I'm a little concerned."

"Really, what's the matter?"

"She's started calling the Psychic Hot Line. Would you believe that she's convinced that those people can predict the future?"

How silly is that? The executive believes that his wife might need help because she is putting faith in the Psychic Hot Line, and yet he expects his managers to predict the future with accuracy! How ridiculous to place all of the organization's plans on those predictions ... and then, punish the ones who fail at the task.

De Geus says it very clearly, "The future cannot be predicted. But, even if it could, we would not dare to act on that prediction."

The only hope we have as managers depending on the predictions of their staff is that there will be "no surprises."

Some managers actually make that a mantra, "No Surprises." They don't want their plans disrupted, they don't want to go to the boss, or the Board of Directors, or the investors, and tell them that things didn't work out the way they planned.

What a delusion! When have things ever worked out as planned? We have to assume that there will always be surprises, and we have to be prepared to handle them, even take advantage of them.

So there you go, two of the most important reasons for doing away with the current budgeting process. You can't predict the future with any accuracy, and the current budgeting/reporting process doesn't work when times are turbulent. Probably won't work in the 21st Century when things are changing at an exponential rate.

What else is wrong with this process?

- It takes up an inordinate amount of management time.

I am the inventor of computer solitaire; that is, the first producer/game designer to create the first commercial version of solitaire on the computer. Most often, when someone discovers this, the first response I get is, "So you're the guy who is single-handedly responsible for the greatest loss of personal productivity in the world?" My answer is, "Regarding the loss of productivity, solitaire doesn't hold a match to the budgeting process." Their usual response, "You can say that again!"

I estimate that the average manager in most large companies has a major focus on budgets about two months out of every quarter, with one month off before the process begins again. For a process that is basically broken, that's a lot of wasted effort, wasted productivity.

- It lacks flexibility and therefore hampers efforts to meet unforeseen needs; stifles creativity.

Often in companies a manager will have an unforeseen need; if it isn't in the budget … forget it. You are committed to that budget, there are precious few exceptions.

This reminds me of another story about Joe Kipper.

Streamlining the Budgeting Process

I had gone over to his store to see if he could discuss an issue with me. His office manager told me he was, "Out watching a guy paint our building."

I went out and walked around the store. Sure enough, there he was watching this guy paint the building.

"Joe, what are you doing out here watching your building being painted?"

"I'm not," he responded.

"What do you mean?" I asked.

"It's simple, I'm not watching my building being painted."

"If you're not watching your building being painted, what are you doing?"

"I'm watching some file cabinets being delivered."

Frankly, I was confused.

He could see that on my face, so he said, "I didn't have any money in the budget to get the building painted, but I did have a lot for new file cabinets. I didn't need the file cabinets, but I sure needed the building painted ... they wouldn't let me paint the building ... so I told them I was buying file cabinets."

This is one of the characteristics of the Effective Leader, being willing to do what you need to do, to get the job done. But, there's another lesson here. The company's budgeting process was forcing Joe to lie, to manipulate the system in unacceptable ways to accomplish what he needed to accomplish. He was actually being put at risk, if his boss, or the financial VP, decided that he was untrustworthy.

And, none of this was Joe's fault. It was the fault of an inflexible system that fostered unreasonable attitudes in the Sears' financial organization, an organization that should be serving the stores, not hampering their efforts to succeed.

- It encourages the spending of money on unneeded materials and resources.

The opposite situation can be found at year's end, when you find managers scurrying around trying to spend up to their budget limits, because they know if they don't, they won't have the budget next year. This type of attitude, again fostered by an unworkable system, is unacceptable and will be extremely counterproductive in this century.

- It moves control and authority up in the organization, while moving responsibility down.

In my opinion, when thinking about how leadership will be impacted by the coming changes, this is the worst effect of all. Leaders, to be effective, must have authority in line with their responsibility, must be able to make the decisions needed to get the task accomplished. Anything that takes away reasonable authority is wrong; even stronger: *it is a sin within the context of the organization.*

I can hear you asking, "So, I'm convinced the cure seems to be worse than the disease ... what's the alternative?"

This question was a problem for me, just like it appeared to be for de Geus. However, I'm more radical than he is, and I would have suggested just tossing the whole thing out and doing without any process at all. Some very successful companies get along fine without a budgeting process that involves every manager in the organization. When I was at Mervyn's, before they were purchased by Dayton Hudson, we didn't have a budgeting process. I'm sure top management put some kind of figures together for the investors, but middle management was never involved.

We did what we had to do to get the job done and nothing more. Probably, if I had gone overboard, I would have been asked to be more careful ... but the culture in the company included an attitude of being fiscally conservative. I had bought into it, nobody had to tell me to be careful and only spend what I needed to.

Later, after we were purchased by Dayton Hudson, budgeting was instituted. The only difference I could see was that I couldn't accomplish as much because so much time was spent on budgets ... I still did what I had to do—whether it was in the budget or not.

This is where effective fiscal responsibility begins, at the heart and soul of the company, in the culture; in the attitude that you only spend what is needed, for what will pay back a benefit equal to or greater than the cost. This attitude is one of responsibility and trust, not control and manipulation.

This is the mature attitude that will be the hallmark of the successful organization in the Twenty-first Century. Let me state this very clearly: Enlightened Companies will by nature be fiscally conservative, this attitude will be an essential part of the corporate culture, of the company's mission statement. This is an essential characteristic, if companies expect to survive during times of rapid change.

I was struggling to come up with a practical solution, when one dropped into my lap.

I was sharing with my Managerial Communications class the issue of budgeting and the problems that are caused by the process, when one of my students, Sheryl Mackey, said, "We didn't use budgets when I worked as an accountant with Nippon Broadcasting News out of New York City."

Nippon Broadcasting is one of the largest broadcasting networks in the world and the major network in the Japanese market.

"You didn't. Why not?"

"The decision was made that the news was too unpredictable, and that budgets were therefore worthless. Worse than that, it was felt they would get in the way of getting the news."

She went on to explain what they did as an alternative, and I quickly realized that I had my answer. Here was a perfectly good alternative that met basic needs without hampering the flexibility needed to be a leading news organization. Additionally, the solution takes very little management time, freeing management up to focus on getting the job done.

What I liked about what the Nippon Broadcasting News approach, is that they don't attempt to predict the future. Managers are responsible for putting together a sheet of financial norms that reflect expenditures if everything goes normally. Then each manager has the responsibility to spend what's needed to get the job done, but no more. The company is reasonable in these instances, which happen all of the time, but they expect the manager to be reasonable also; much like my early experience at Mervyn's. Everybody wins, and nobody is involved in a major effort that, when finished, isn't worth the paper it's printed on.

Instead of being responsible for predicting the future needs of their areas of responsibility, managers will be responsible for developing a series of financial norms that describe fully how much they will spend, if things continue as they are currently. Adjustments to these norms are submitted as things change and, under most conditions, quickly authorized. The manager is responsible for making both positive and negative adjustments (surprises can also result in a necessity to lower the norms). This process will have a significant positive effect in the following areas:

- It doesn't ask the manager to predict the future and then punish him when he is wrong.
- It provides a way of dealing with surprises (turbulent times).
- It doesn't take an inordinate amount of management time. In fact, most of the time will be spent considering how to handle surprises. This is how the available time should be spent.
- It is extremely flexible, designed to handle unforeseen situations.
- It eliminates the tendency to spend to maintain budgets at their current level.
- It gives managers the authority equal to their responsibilities. The process believes, has faith, that managers will make the right decisions, that they understand and support the company's commitment to being fiscally conservative.

One final thought on this subject. I know that a major reason this process exists is for the investor community,

Streamlining the Budgeting Process

a group only interested in return on investment. They are convinced that companies ought to be able to predict the future (this group probably does go to the Physic Hot Line from time to time). And they not only want to know those predictions, but they want to punish any company whose predictions do not come true.

This attitude is dramatically affecting those companies' abilities to get their real work done, the work of building a company that will be successful over the long haul. So, in essence, one of the biggest problems regarding productivity lies in the laps of the investors themselves and their demands. This means that investors share the responsibility when a company fails.

It had been my hope that companies all over the world would see the necessity of developing a system of financial responsibility along these lines. A process where the budgeting was reasonable and quarterly reporting was eliminated.

However, since I originally wrote this essay, the governments of the world have decided to do exactly the opposite: require that all public companies in the world move to quarterly reporting and maintain comprehensive budgets that can be used to determine the worth of a company. This is a great disaster that only makes it more difficult to be effective. As I just said, in many ways, the investor community is its own worse enemy.

11
Defeating Elitism

The truth that I have seen is that there is no additional knowledge, no greater intelligence, and no more wisdom, within the community of those in charge, than there is in the community of the rest of us. The only characteristic that the elitists share is an arrogant ignorance that can't be overcome.

In the beginning, the elite amongst us were those with either the power to destroy us, or those with control over knowledge. Knowledge, too, was power. However, the Internet has changed all that. As knowledge and the ability to gain needed skills becomes available to all, we, the organizations of the world, will find our leaders from every walk of life, from every race and persuasion. This will become one of management's greatest challenges; to be able to accept and embrace the leaders within their organizations as they make themselves known.

The elitism born during the time when knowledge, information, and critical skills were only available to the wealthy will prove to be counterproductive in this century. Knowledge will be available to all with the moxie to gain it. One of the reasons for this is that the signs of the elitist are going away. We used to be able to tell who was "important" by the clothes they wore, or the car they drove, maybe the school they went to, or the clubs they belonged to. You can't depend on that anymore.

Business will no longer be able to depend upon resumes', educational history, or business background to select the "best" candidates for the job, the best people to join the team. Leaders will develop additional techniques to determine which individuals are best suited to become members of the team. Personal characteristics like ambition, integrity, smarts, moxie, energy level, and knowledge and skills (including those that are self-taught) will become more important, while business and educational background will be considered only when it can be determined that that particular experience is valuable under specific conditions.

With a rapid rate of change, they cannot afford to have incompetent members holding back the potential of the team, ineffective management controlling their destinies. The arrogant ignorance so prevalent today will go away as the power of Effective Leadership wins the race time and time again.

Elitism impacts every critical area within the company. It encourages the hiring of individuals who may not be the best for accomplishing the company's goals. It fosters a dependency on outside consultants with "proper credentials" even at those times when everything that is needed and/or known exists within the team that is already in place. It leads to a lack of respect in, and trust of, the team, with a potential of lowering morale and crippling motivation. And it robs individuals of their opportunity to "make a difference."

Elitism is especially dangerous during times of rapid change when the best and most effective of our leaders must be in place. Elitism will simply have no place in the Enlightened Company.

Today, if you're working in a software development company and somebody shows up in a suit and tie, they are immediately suspected as being the enemy. They gain no advantages, in fact they gain no respect. The question that is asked is, "Who are the suits?" and the tone of voice is not kindly.

I was told when I started speaking that I had to present the appearance of someone who knew more than my audience, someone who was superior to them. It's believed that this is the only way to get credibility, gain the respect needed before the audience will listen to you seriously. I'm here to tell you that those times are changing.

Students today do not give respect to suits, they respect those that teach them something valuable, something that will increase their knowledge and potential.

They respect teachers who inspire, who are committed to their work, who believe in what they have to say, and who know how to say it. They also respect teachers who can listen to disagreement, admit when they are wrong, and change when the need for change is apparent. You gain their respect by who you are, not by what you are wearing.

It isn't only software developers who are tired of suits and ties. A friend of mine Chihoe Hahn, was a young attorney at Wilson Sonsini, a very successful law firm in Palo Alto, California. They had outgrown their present facilities and wanted the junior attorneys to move to offices across the street. Chihoe was one of the ones they asked to move.

Chihoe, like me, is a maverick, and he refused to go. In fact, he organized all of the young attorneys, so that they presented a united front. He believed that if they moved, they would be out-of-sight and that this would have a negative impact on their careers. The firm's senior partners came to Chihoe and offered them a yogurt machine if they would move.

"We're not going to jeopardize our careers for a yogurt machine."

"How about we throw in Starbucks Coffee? Always a fresh pot, all day long ..."

It dawned on Chihoe that they were desperate. They had to get them over to the new building. He wondered how far they would go.

"I'll tell you what, we'll go to the new building if we can have casual Friday's every day of the week. You're not planning on having clients to the new building, so there's no reason why we have to dress up everyday."

The senior partners agreed. No more suits, no more ties, a time for celebration, right? It wasn't that easy. At the start only one other junior lawyer had the courage to come to work everyday in casual clothes. In fact, Chihoe and the other guy would call each other every night and commit to wearing casual clothes the next day. He wondered how long they could keep this up.

After a couple of weeks, a few others started being causal more often and then it was a done thing; the satellite office of Wilson Sonsini was 100 percent casual everyday of the week. Then the lawyers in the main office began to get jealous. "If those kids can dress casual so can I!" Before you knew it, Wilson Sonsini was 100 percent casual everyday of the week.

Then other law firms noticed what Wilson Sonsini was doing. Senior partners in those firms were meeting behind closed doors, "This is a calculated event. Wilson Sonsini has figured out that the Silicon Valley clients are more comfortable in casual clothes. They're trying to take away our business by appearing more friendly. We've got to do something!" So, in order to be competitive, the rest of the Silicon Valley law firms went 100 percent casual everyday of the week.

Defeating Elitism

Their professionalism, their credibility, wasn't hurt at all; in fact, it was probably helped. There's nothing more scary than someone who's both a lawyer and a suit!

We are the generation that is experiencing the greatest changes ever experienced by the human race, change that only began accelerating at this great pace within the past 100 years. And, the pace is increasing at a rate that is truly beyond the ability of any of us to imagine. Only Enlightened Companies and Effective Leaders will be able to handle it, to adapt. Many of our current companies and managers will be overwhelmed. Is your company enlightened? Are you on the path to Effective Leadership?

12
Business Ethics in Crisis

Without a healthy community, society, or world, where will companies find their resources, human and natural? Without healthy communities, societies, world, where will companies find their customers?

At Texas State and Franklin Universities, we take ethics and stakeholders seriously, with sections of almost every class devoted to discussions along this line. From these discussions, the confusion around this issue becomes apparent. And that confusion is only a reflection of the same confusion that is felt throughout our society. For example, this was a complaint from an executive taking my business ethics course:

> Yes, we need to care about the environment, but how do we do that and remain faithful to our fiduciary responsibility to our shareholders? They expect to see gains with every quarterly report; often what we need to do to accomplish this, is in direct opposition to what is right for the community, our suppliers, even our employees.

Allan Sloan, *Newsweek*'s Wall Street Editor, tells us that greed isn't good, that the conventional wisdom of the past few years, "… the pursuit of self-interest (results in) … public benefit …" is not right and leads to all kinds of activities that are not in the public's best interest.

Of course, I support this position; but I say it doesn't go far enough, doesn't identify the core issues.

This is not a simple problem that can be solved by convincing people that their interests are better served when we and our companies operate ethically, with a strong sense of the public good. The solutions begin to surface only after we understand more clearly how we got to where we are, and where that is.

Before we look at that, I want to state clearly that I am a staunch supporter of capitalism, an Independent who tends to vote Republican. However, the system is badly broken, in many ways and events such as the Enron scandal are only symptoms of a disease that goes much deeper.

Initially, our capitalistic society had a chance to operate freely with little government intervention. The country was large and communication was slow; there was no such thing as a national corporation, let alone a global one. Additionally, if we, as businesspeople, were wronged, cheated out of our just due, we understood that the courts wouldn't be able to help much. So we took matters into our own hands; we placed ads in the papers warning people not to do business with so-and-so, "He cheated me!"

But even then, capitalism (the free market) had one major flaw; without regulation it was a one-winner system. This meant that many communities were held under the sway of a single person, beholden to them in all ways; hence, "It's a Wonderful Life." In many ways, capitalism, allowed free reign, is much like a game of Monopoly; in the end, one person ends up with all the property and all the money.

Then communications improved, and travel from one end of the continent to the other also improved dramatically. We began to see national companies with powerful self-interests having impact on local economies, often negative impact. Finally, it became necessary for government to step in, and the Taft-Hartley bill was passed, making it illegal for a monopoly to take advantage of its situation in ways detrimental to the public good. The problem: detrimental to the public good was defined in a limited fashion, basically meaning higher prices.

Along the way, other issues surfaced, that play a major role in the current situation. They are:

- Business started operating on a quarterly reporting basis.
- The fiduciary responsibility of company directors was declared to be only to shareholders.
- Our legal system failed to protect small businesses from illegal and/or unethical practices.
- There was, and continues to be, lack of understanding of the true danger of monopoly to our way of life.

Let's look at each of these more closely.

Business started operating on a quarterly reporting basis.

I'm not going to pretend that I fully understand why this practice began, but I will say this: It is a system designed for the convenience and possibly the protection of shareholders. The problem: This very activity, ostensibly designed to "protect" shareholders, is probably the greatest deterrent to corporate success in existence.

Intelligent people know that three months isn't a long enough time span for anything to happen. Effectively it's an attempt at micro-managing from the highest possible level. And it's a terrible waste of management energy.

But worse than that, it forces companies to focus on the short-term; with the result that management is almost legally bound to activities that could be disastrous in the long run. One result of this focus can be quoted right out of Sloan's article; "All sorts of companies are trying to duck U.S. taxes by incorporating offshore. They're doing it in the name of staying competitive in a global economy."

And they're doing it as a result of their fiduciary responsibility to shareholders, to the detriment of our society and ultimately to our way of life.

The solution: Semi-annual reporting.

The fiduciary responsibility of company directors was declared to be only to shareholders.

The next obstacle to change is a lack of understanding regarding the fiscal responsibility of companies to their stakeholders. Currently companies only recognize their fiscal responsibility to shareholders. All other responsibilities are lesser in nature and therefore suffer during times when it is necessary to "cut back." This situation will have to change in the future.

As long as these laws are in place, we can talk ethics and morality till we're blue in the face; but we will have no more impact on the situation than Demosthenes had on the tide, speaking loudly at it with his mouth full of pebbles.

With the potential increasing to do great harm to our society, even to the world, organizations are going to have to take the longer view, accept their fiscal responsibility not only to their shareholders, but also to their employees, suppliers, customers, communities, even to the world. Businesses will need to learn to accept fully the proposition that their long-term health is dependent on the health of the communities that they serve. This is the only way that they can be assured of the resources and customers that they will need to survive in a rapidly changing world.

I have been blessed to have been involved with companies that understood the importance of giving back to their communities. Mervyn's takes this responsibility very seriously, giving back a full 5 percent of the net profits. In fact, Merv Morris, Mervyn's founder, was also the founder of the San Francisco Bay Area's 5 Percent Club.

There is no doubt that business is on the right track, is beginning to understand the importance of "giving back." In the next century companies will move beyond this altruistic motive, a commendable motive, but one that loses its strength when companies are going through rough times, looking for ways to cut costs. We are close to a full understanding of the importance of individual companies working in their communities, the society, the world, to assure that each is socio-economically healthy, because healthy communities are essential to the long-term success of our businesses. The fact that almost every company being formed today is multi-national (this is another benefit of the World Wide Web) will surely help to bring this realization to the forefront.

Let me state it categorically: Businesses have a fiscal responsibility not only to their shareholders, but to their members, suppliers, customers, communities, society, and the world. This responsibility is critical to the long-term success of the business. Any company not accepting this responsibility is a parasite feeding off of the efforts of those around it.

I remember discussing this with an old friend of mine years ago. I was disappointed at the attitude of many of the high technology companies who seemed to take the stance that they were already saving the world with their technology; therefore, they didn't need to do anything else.

Kerry, my friend, was a major executive with a New York bank, so I knew that he was well aware of businesses fiscal responsibility. But he was also extremely active in his church and had a strong sense of community. I expected him to support my position. He didn't. Instead he wanted me to understand that, "It is very clear that management's only fiscal responsibility is to shareholders. They are legally libel if they spend profits in ways that do not benefit the shareholders."

This attitude is pervasive and dangerous to the long-term health of the company, of our society. As I stated earlier, without healthy communities, organizations cannot survive. In the, hopefully, transitory age of the quick-buck, make-a-killing and then cut-and-run attitude, there is very little chance to think much beyond the next quarter. This kind of company can't survive for the long haul. People want to be involved in things that last, efforts they can be proud of, visions they can be committed to. This is how you attract the best people; this is critical to understand, if you want to build teams that accomplish the impossible, or close to it.

There is no doubt that any responsibility focused entirely on shareholders (especially when coupled with quarterly reporting) and ignoring other stakeholders, is short-term focused and ultimately detrimental to all stakeholders, including shareholders. In many ways, investors are their own worse enemy.

In my opinion, the opposite is also true; if director fiduciary responsibility was shared equally with all stakeholders, companies would automatically become long-term focused, boards of directors would be legally bound to consider the ultimate impact of their decisions, and long-range planning would become a natural part of running a business, instead of an activity that only gets lip service.

The solution: Fiduciary responsibility expanded to include all stakeholders.

Our legal system failed to protect small businesses from illegal and/or unethical practices.

Finally, throughout this chapter I have shared with you my strong belief that trust, trustworthiness, respectability, a strong sense of business ethics and morality, will be an essential characteristic of the Enlightened Company. Companies that cannot be trusted, cannot be depended on, will not succeed in a rapidly changing environment where your first choice for a vendor or strategic partnership may be your last choice. There is another catalyst that will help to bring about this desired change, the failure of the legal process in America.

Even though the legal process is failing the vast majority of citizens in the United States, this fact is only slowly becoming apparent to most of us. We still want to believe in justice, we want to believe that we will be judged "innocent until proven guilty;" we want to believe that if we are honestly wronged, that wrong will be righted. The fact that this is not true in both the criminal and civil arenas is still not widely known, nor completely accepted.

Initially, I saw the failure of the legal process as a great negative, one that allowed the wealthy guilty to get off Scot-free, even dramatically increase their position, while causing the vast majority of accused, innocent Americans unbelievable emotional suffering and financial, if not actual physical, hardship. Most small businesspeople and their attorneys know that there is no justice and no recourse through the courts if they are cheated out of tens of thousands of dollars, or attacked by a major corporation.

I have come to see this situation as a blessing that will bring about a resurgence of honest, ethical business practices. Business relationships will become more dependable as more and more companies realize that they can only afford to do business with people and organizations they trust.

In fact, I can see the Internet providing a tremendous opportunity for this to happen naturally. We are already seeing it in the auction sites, where the integrity, dependability, and trustworthiness of participants is up there and available for all to see. The process isn't perfect, for example, untrustworthy individuals can just sign on using different names from different computers.

But this situation is only a result of the infancy of the industry and it will ultimately be perfected. People will quickly discover that they must act ethically with responsibility, or lose their credibility, their ability to do business on the site.

Imagine if this process were extended to all business of all kinds, if you could check the reputation of the company, of the individual, before you did business with them ... imagine how that would change the shape of business today.

The impact of trust and trustworthy relationships on the face of business for the Twenty-first Century cannot be over-stated, nor over-emphasized.

The legal process within capitalism was never very good at protecting small business and small businesspeople (the true backbone of our society) from attack by others, be it someone's refusal to pay a significant bill, or a major attack from big business, with its power to easily drive any small business into bankruptcy.

This situation has not improved; if anything it has gotten worse. The minimum costs for defending your rights as a businessperson in Federal Court is approximately $70,000; even if the law is on your side and your position is morally and ethically sound, you have no guarantee of winning. Most of the time, the winner is the one with the best attorneys and the most money ... justice simply doesn't exist.

Even worse, if you do win, it is still extremely difficult to claim your legal fees as part of your losses; you can be right and win, and still be forced into bankruptcy.

The result is a feeling among the unethical and immoral that they can do what they want, take what they want, destroy whomever they want; and the penalty, if they lose, is a small one to pay—just close up shop and go looking for another sucker.

I have no faith that the legal process will ever be fixed. Why? First, it's a complex issue with many agendas. And, because we have to depend on lawyers to do the fixing, and the conflict of interest is much too apparent. (If I had my way, I'd make it illegal for lawyers to make laws. It's all too obvious to me that we've hired the wolf to protect the chicken coop.)

The solution: Create a court between small claims and Federal court that allows for claims of up to $100,000 and doesn't require tens of thousands of dollars to access. One way this could be accomplished is to have it work much like arbitration, with a strong focus on what's right, legally and ethically; where the power of the lawyer is not greater than the power of the truth.

Sounds naïve doesn't it ... Well, I do have one other alternative.

Alternative solution: Go back to an honor system where individuals are free to announce to the public, who has cheated them, and how they did it. Then refuse to do business with those who can't be trusted. Likewise, see it as a responsibility to share the good news when it is discovered that we're doing business with an honorable person or business.

There was, and continues to be, *lack* of understanding of the true danger of monopoly to our way of life.

When the Taft-Hartley Bill was being passed, it was obvious that consumers, and therefore the public good, were being harmed, so the bill focused on that. Today it is often not obvious that this is happening. Therefore, confusion reigns. Monopolies argue that they are good for society, "We bring about a consistency that's essential to moving forward in ways that we must, if we are to succeed as a society." And many, who truly have the public good in mind, support them in this contention.

To begin to understand this problem is to first accept that capitalism is essentially a "one-winner" system; if left unchecked, one person or one company, would ultimately be in control of everything and everybody. Government must understand its responsibility to make sure this doesn't happen. Congress must step forward and say, "Monopolies are not ultimately good; they are always a danger that must be watched carefully." It's time to take the long-range view and realize that the dangers of monopoly go way beyond the consumer's pocket book, reaching into the very heart of our way of life.

Capitalism flourishes when individuals have a strong sense that their efforts will be rewarded, that they, too, have a chance at the brass ring, maybe even a gold one. Monopolies, by their very nature, hoard all of the rings; do everything in their power to make sure that nothing interferes with their current business or future goals.

Even worse, they begin to believe their own press, "What's good for GM, is good for the nation." Or, "What's good for Microsoft, is good for the world." If we knew how many businesses Microsoft has stolen the passion from, even forced into bankruptcy, we might be appalled. And yet, many hold Microsoft up as a standard that we should all emulate.

In many ways Microsoft is the most powerful monopoly the world has ever seen. Do they harm consumers, where it currently counts, in the pocketbook? As the U.S. government and the states are finding out, that's difficult to prove (one judge had it right).

Is the Microsoft monopoly an ultimate danger to our way of life? I have no doubt as to the answer to this question: Absolutely! Left unchecked, Bill Gates would literally try to take over the world, probably rationalizing that it's the best path to world peace.

The solution: Pass up-to-date laws that recognize the true nature of monopolies and take into account the long-range public good.

13

Embracing the Unexpected

The Effective Leader seems to accomplish the impossible, creates the "luck" that makes doing difficult things seem easy. I call this "Embracing the Unexpected." This is not some esoteric concept dependent on faith and/or belief to work. This is how things really work.

Earlier I defined the Effective Leader as someone who,

> ... gets the job done in ways that increases the competence, confidence, and potential of the individual team members, while building a team that can accomplish the seemingly impossible (at least the extremely difficult); all to the long-term benefit of the organization."

Likewise I listed sixteen characteristics of individuals who are, or have the potential to be, Effective Leaders. Of those sixteen characteristics, five relate directly to the ability to take advantage of surprises, to use unexpected events to accomplish what needs to get done even more effectively ... in fact, making it seem almost easy and therefore appearing to be, "just lucky." The five characteristics are:

- Highly intuitive and not afraid to trust those feelings.
- Genuinely curious about how others think and the ideas/opinions they have.
- Sensitive to cues concerning potential problems and opportunities.
- Willing to embrace the unexpected and the resulting possibilities.
- Use of centering techniques for relaxation and heightened awareness.

Let's start by defining these three terms:

Coincidence: Related events that happen by accident, the coming together of which usually results in a positive outcome for the involved individual or group.

Synchronicity: Related events that could not conceivably have happened by accident, but which cannot be explained logically, the coming together of which usually results in a positive outcome for the involved individual or group.

Serendipity: The achieving of a significant goal that was different from the one originally planned for.

While we all understand what is meant by the term coincidence, there is less understanding of its sibling, synchronicity. Maybe an example will help.

I was doing a term paper on Astral Projection for an eastern philosophy class I was taking at San Jose State University back in 1971. I was attending San Jose State part time, getting my BA, while working full time managing a retail store in Mountain View, California.

Most of the books I was able to obtain on the subject referred to three books that appeared to be the foundation for the modern study of the phenomena. I could not find any of these books in any library in the San Jose Metropolitan area (the Silicon Valley). This included the Stanford Library, which is one of the most complete in the nation in psychology and parapsychology. (Stanford was originally founded as a University dedicated to the study of these two areas of knowledge.)

After doing all I could to find the books, I released it to the "higher forces" and forgot about it.

The very next day, while at work, I had a strong feeling that I should to go to the Los Altos library and obtain a library card. I tend to honor these intuitive feelings whenever possible, and this day was no exception.

Fifteen minutes later, I was seated next to one of the librarians and she was filling out my library card application. I happened to notice a young man enter the library and then precede to each of the Los Altos librarians and show them a piece of paper that he was carrying.

Embracing the Unexpected

He seemed very proud of something and I was curious to know what it was.

When he approached us, I said, "Something pretty good must have just happened."

"I got a letter from the Dali Lama! See!" and he handed it to me.

It was a letter from the Dali Lama all right. It was essentially affirming the course that he had planned for this life and the success with which he was achieving his goals, and encouraged him to continue with his plans.

"Wow! This is real neat." I said. He smiled happily and took his letter back. I continued, "Do you study eastern philosophy?"

"Yes, I've studied it for years," he replied.

"Does your family study it too?" I asked.

"Oh yes, my dad has a great library with tons of books on everything to do with eastern philosophy."

"I've been looking for three books that I can't find anywhere ... do you think your dad would let me check out his library?" I asked.

"No problem, just give him a call," he replied.

I got his phone number and called his dad as soon as I got back to work. He was a true gentleman, not only offering me a look at his library, but inviting me to have a bowl of soup with them that night after work.

"But, I don't get off until after the store closes ... I wouldn't be to your home until ten o'clock or later," I said.

"That'll work out fine. See you then." And that was that.

Later that night after a nice conversation over a bowl of lentil soup, he said, "Well let's see if I've got those books you're looking for."

We got up from the table and went to another room with bookcases on every wall. Like a moth to the flame, I immediately walked across the room, and there, directly in front of me, were the four books I had been looking for, all together on the same shelf.

This event goes beyond coincidence. If I had run into the young man while reading about the phenomena, and discovered that his dad had books on the subject ... that would have been a coincidence. This kind of coincidence happens to me fairly often, because I am so passionate about what I am doing, reading, studying at the moment, that I tend to share it with anyone I meet.

So, the chances I will share it with someone who has a like interest, or who knows someone who does is fairly likely ... but, regardless, it is still coincidence. What makes this example different is that the coincidental event came exactly when I needed it to, as a result of my request when I released it to the "higher forces." It was as if the event had been planned so that I would achieve the goal that I could not achieve through my own resources. This is *synchronicity*.

Serendipity is something else entirely. This term comes from the legend of the "Princes of Serendip," a mythical land that some say is the modern day Sri Lanka. These two Princes would make marvelous plans to accomplish wonderful things, set out on the journeys that would lead to their goals ... and, because of events beyond their control, never accomplish what they set out to, but always accomplish something even more wonderful.

It is critical to note that the Princes always had a goal in mind, and always made intricate and complete plans to accomplish it. I have heard people say that they aren't making any plans at all, that they are experiencing their lives as a series of serendipitous events.

They are wrong. They are not experiencing serendipity, they are experiencing randomness. Serendipity involves planning, followed by events beyond your control, and the openness to take advantage of the unexpected, ultimately accomplishing something beyond what you originally believed possible. The world of science, of business too, is filled with stories of serendipitous events, of times when what was accomplished was very different and much better than the original goal.

Now let's explore the five characteristics I listed above to try and discover the role they play in helping you take advantage of coincidence, synchronicity, and serendipity.

- Highly intuitive and not afraid to trust those feelings.

In chapter 2 I explained in some depth where intuition comes from, how it's a natural process of the mind and not, necessarily, associated with some mystical force.

This means that human intuition isn't some esoteric process involving something outside of ourselves; it's a natural process designed and honed to help us interact successfully with an ever changing environment.

- Genuinely curious about how others think and the ideas/opinions they have.

A natural curiosity is one of the most critical characteristics needed to lead a successful, effective life. And this will be equally enhancing to your personal and business relationships. With an active curiosity, a very real interest in why things happen, why people think and/or feel the way they do—you don't need a course in active listening.

As a leader, your most important responsibilities include solving problems and handling conflict resolution. The first step in problem solving and conflict resolution is to gain understanding, and you can only gain understanding by encouraging communication and then listening carefully with curiosity from the heart, being fully open, to what you are being told/shown.

While there have been thousands of classes teaching effective techniques for accomplishing this, none of these techniques are worth the time and money spent on them, unless you have an attitude that springs from a real internal need to know what's happening ... a real curiosity about what is going on, why he or she feels, or is thinking the way they are.

    These conversations lead to effective decision making, lead to knowledge and information that you didn't have before. They provide your brain with the data it needs to make your intuitive thought processes work better. You don't need to know why any single piece of information is valuable; your brain will take care of that. You just need to be as sure as possible that it has as much data as possible.

    If you look at it this way, you will understand why these conversations are so important. You don't have them because you want the person to believe you care, you have them because you will make better decisions if you do. The fact that the person will believe you care what they think is a great side benefit; that happens because you do care.

    When you patronize to make someone feel better, they see through it, and they aren't satisfied. When you communicate for yourself, so you will learn more and make better decisions, they sense your sincerity and they are satisfied.

- Sensitive to cues concerning potential problems and opportunities.

I call this "cue sensitivity" and it is one of the most important skills for an Effective Leader to develop. While your brain will use these subtle cues to provide you with intuitive knowledge about a situation that needs your attention, it is possible to learn to become consciously aware of them when they happen.

When you become aware of these cues, you are then able to respond to them appropriately. This early appropriate response will very often eliminate any possibility that the situation will mature and cause you real problems. The use of this skill is why Effective Leaders tend to avoid "Murphy's Law" (Whatever can go wrong, will go wrong), why they seem so lucky.

Let me give you an example. Let's say you meet someone in the hall and they say, "Jim isn't looking happy, do you have any idea what's wrong?"

You respond, "Not really, but we've all been under a lot of pressure lately."

"That's probably it."

As you walk away you think, *Maybe I should talk to Jim ... just in case.* That's the last thought you have on the subject until three months later when Jim resigns in anger, fed up with the lack of cooperation he is getting from a critical, internal vendor.

    Now you're thinking, *I knew I should have talked to him.* The problem is, it's too late; Jim's gone and the critical project he was managing is in jeopardy ... there's no doubt, anything that can go wrong, will.

It is cue sensitivity that enables you to head off a problem when it is still manageable; without this skill you drift from one major disaster to another, wondering why this is always happening to you ... why you're so unlucky. All it took was being sensitive to this cue, followed by a simple conversation with Jim, "What's going on, everything okay?"

    If you're open to his issues, listen carefully, and are curious enough to dig out the real problem, there's a great chance that you'll be able to solve the problem, heading off the disaster waiting around the corner.

    In addition to helping you head off problems, cue sensitivity can also help you take advantage of surprise developments. This is the difference between being aware of what's happening and ignoring those developments that don't fit your paradigm.

    The history of science is full of instances where one individual looked at what was happening, noticed it, and wondered why. My favorite example is the physicist,

Richard Feynman, who was watching a magician twirl plates on the end of stick and wondered about the unusual wave pattern the edge of the plate was exhibiting ... this moment of cue sensitivity, coupled with a strong curiosity lead him to the Nobel Prize and a reputation as one of the finest modern physicists.

Again, your brain will analyze this data, and then it will feed you the information as an intuitive thought that you can either act on or ignore. But, you can learn to be sensitive to these cues, to become cue sensitive. When you do, you are on your way to becoming a truly Effective Leader.

- Willing to embrace the unexpected and the resulting possibilities.

Why are we so afraid of the unexpected, so worried about being surprised? This fear really has no place in life, because life is full of surprises and you ignore that fact at your detriment. Surprises are just as often positive as they are negative, just as often lead to better results, as they are obstacles to progress. In fact, very often the surprises that seem negative are only opportunities to improve the product or the process ... and are, therefore, extremely positive in the long run.

The willingness to embrace surprises is found in your belief that you don't have all the answers. It is fostered by a lack of arrogant ego that enables your natural curiosity to wonder what's going on and what it could lead to.

It only happens when you are freed from the belief that the plan is all important, that you must accomplish not only what you have set out to, but you must accomplish it in the way that you planned to. Too often, we let the pert chart run the project, when it should be serving it. This attitude does not lead to a willingness to embrace the unexpected.

This willingness becomes easier as you learn to trust your intuition, as your curiosity uncovers unknown potentials, and as you become sensitive to the cues in your environment that are hinting at the possibilities you haven't even considered.

- Use of centering techniques for relaxation and heightened awareness.

Much of what I have been talking about depends upon a healthy relationship with your subconscious mind, that part of you that serves you in so many ways. It warns you, through intuition, that something is happening of which you need to be aware. And it protects your ego so that you won't be surprised by learning that your perception of reality is not the truth. I had an interesting experience once that demonstrated the power of my subconscious to let me see only what I believed to be true.

I was the store operations manager for a Mervyn's in Mountain View, California and occasionally on Sundays I was the acting store manager. When I had this duty, I took it upon myself to do one final tour of the store just before opening. I did this for two reasons; first, I wanted to say hello to everyone and let them know I appreciated their coming in to work on a Sunday, and second, I wanted to check my coverage, see where I might have some extra help, if I ran into trouble elsewhere.

The Mountain View store had an unusual layout—the Boys/Toys and Domestics Departments were off on a separate wing, while the rest of the departments were fairly well centralized. As I was getting ready to, "take the long jaunt," I ran into the Domestics saleswoman and she let me know that she was all alone that day and would need break coverage. Since I knew that the Boys/Toys Department had only one person, I decided not to take the jaunt until a little later, so I started walking back to my office. I quickly realized that there was another reason to visit the department, to greet and thank the person there. So I turned around and walked over to Boys/Toys.

As I walked into the department I saw Jane standing at the register folding t-shirts. I walked up to her and said, "Good morning, Jane, I see you're all alone in the department this morning."

Embracing the Unexpected

And that was when it happened.

There was kind of a roar in my ears and a black curtain descended over my vision.

Through the roar, I heard a woman's voice saying, "Mr. Fregger, I'm here too!"

As the voice spoke, the curtain rose and I could see Betty standing there talking to me, standing right next to Jane. I swear that she wasn't there a moment before, but there she was now. What a shocking experience! I quickly made a joke of the situation and went on my way.

What happened is obvious to me now. I *knew* when I walked into the department that there was only one person on duty. My eyes saw the other person clearly; after all, my eyes are a tool that takes in visual data and then sends the information to my brain. My sub-conscious gets the data first and then decides whether or not to transmit it to my conscious mind.

This decision to transmit or not is dependent on lots of things: how important it is, how interesting it is, and whether or not it supports my current belief system.

This is critical to understand. As leaders you have to make judgments continuously which result in decisions,. You depend upon having the information needed to make the best decisions.

However, if you have a judgment that a member of your team is not reliable, there is a strong chance that you will only notice behavior that supports this belief; vice versa, if you believe that a member of your team is exceptional, you will only notice behavior that supports this. It should be obvious that this proclivity is potentially dangerous in both instances.

You can minimize this natural tendency, and get a firmer grip on reality by using techniques that relax and center you. These practices will heighten your awareness and help you develop a healthy relationship with your subconscious mind. Additionally, that healthy relationship with your subconscious mind will enhance your intuitive powers, your cue sensitivity skills, and your creativity—all needed when you are faced with unexpected surprises.

I use both meditation and self-hypnosis to accomplish this end ... others use exercise, quiet contemplation, prayer, etc. What matters is that you take your subconscious seriously; realizing that it is an invaluable partner in your quest to get things done, accomplish the impossible, and other difficult things.

Embracing the Unexpected

14

Our Future in Space

The images that have the greatest effect are those images that are positive, challenging, and goal-oriented. What image is more positive, more challenging, and more goal oriented than the image of the destiny of humanity to discover, explore, and settle the Universe.

Because the vision or "image of the future" works in the present, very soon after a society accepts the vision, changes begin to happen. The vision does not operate in some other dimension, magically changing the present and, therefore, creating the future. No, it works in the attitudes and energy of the individuals holding the vision, changing the way we think, changing the way we work, changing our priorities immediately.

As the vision begins to take hold, decisions are made differently; investments are made in a different way, affecting others who then, also begin to do things differently. Research is begun that will enable the vision to be realized, investments are made in "new" technologies, people are put to work, and discoveries are made; discoveries that had not been originally planned, ways to benefit society and individuals within the society that were not initially considered.

Citizens react to the image of the future as if it were already here and, therefore, act in a purposeful way that is directed toward this future's goals and expectations. It is the vision that puts the population to work, that increases the productivity of its people, and it is the productivity of its people that is the wealth of a society. Many people, even experts, believe that the wealth of a society is tied up in the amount of gold it has, or the amount of natural resources. This belief continues in spite of the many lessons of history that tell us otherwise.

History shows us three types of conditions that provide the type of vision and energy needed to keep the engines of a society running: 1) war, 2) religion, and 3) a new frontier to explore and settle. In addition, there are isolated cases where societies band together with energy and industry because of a common "threat" from without.

In recent history it has been war that has been the most successful in increasing our drive and productivity. For example, the United States entered World War II a poor nation, coming out of the Great Depression. Four and one-half years later, the war was over and the United States emerged as the greatest economic and military power on the Earth. This "wealth" was accomplished without stealing any of the "treasures" of

Europe or gold and diamonds of the world. At the end of the war we did not demand reparation for what the war had cost us. In fact, we immediately began a restoration program to bring those nations back from their personal devastation, spending additional millions in the process. Also at the end of the war most of what we had built during those years was either at the bottom of the ocean or on a battlefield in Europe, and, most sadly of all, many of the strongest men and women of an entire generation were lost in that effort.

And yet, our economic depression was over and we were a rich nation. Why? What changed in those few short years? The answer is simple: our people went to work. The productivity of our people was at an all-time high.

It was our image of ourselves and the impact that image had on the productivity of our people that made the difference. And the difference was felt by our society almost immediately. Only four terrible, war-torn years took us from devastating depression to socio-economic health. We believed, both during and after the war, that we were the greatest nation on the Earth, and that was what it took to bring about that reality.

War and religion (sometimes both together) have often had this effect on societies. War efforts are no longer an acceptable solution to bring about a healthy economy, to support our pursuit of happiness.

The images that have the greatest effect are those images that are positive, challenging, and goal-oriented. Again, what image is more positive, more challenging, and more goal oriented than the image of the destiny of humanity to discover, explore, and settle the Universe. It is up to those of us who have caught this vision to spread this image; our government and institutions cannot lead us on this great adventure into the infinite frontier. They, by their nature, only mirror ideas and feelings that already exist in society.

If we are going to create a new vision, then it must grow from within each of us. It must grow from the fertile soil composed of strong, confident individuals exercising their free voice within the community to work for the future we see as our destiny, to create and share the vision we have of a nation, a world, a Universe that we would like to live in.

It is my belief that if we hope to convince others to accept this immense vision and support the challenge of the "high frontier," then we must share our vision of a future that includes not only humanity in space, but a better life for the majority of people who will choose to remain on the Earth.

It is not enough for us to share the facts and figures that support the space effort, or to remind those, with other priorities, what the space effort has already provided in the area of scientific or technological advancement. The facts and figures that we have are important; they affirm our vision and tell us of our progress toward it. But, until others accept the vision, until their perception is changed, they will not hear and they will not understand. One of the truths of the human condition is that knowledge inconsistent with a present belief system is ignored, and all the proof in the world will not convince the person who isn't able, or doesn't want, to believe.

If this vision is to succeed, we will need the financial and emotional commitment of a significant percentage of society; people who are worried about paying the rent, or wondering if they can afford a new car, or concerned about a loved one ill with a disease or injury.

Those who are speaking out against oil drilling along our coastlines, or in the heart of Alaska, or worried about how modern technology is affecting the climate, changing relationships, destroying the planet, these people are going to have to believe that the space effort is a first-level priority, that their deepest concerns will be aided by our efforts in space, that the Earth will be a better place to live as humanity accepts the destiny we believe in and that, as we turn toward the stars, we will also find solutions to many of the problems that are multiplying on this Earth.

Think about the potential impact on the attitude of people on the Earth toward ecology when the total ecology mindset of the space colonist is fostered in those people still living on the Earth. In space we must conserve; we know that we are living on a spaceship. That attitude of conservation may well spread back to the Earth with people more readily accepting the concept of "Spaceship Earth" with the limited resources that implies.

The most important reason for supporting this vision is the effect that it will have on our image of ourselves and of our future. Let me ask you a question. Would a national goal involving the "high frontier," space industrialization, and ultimately space settlement result in a positive, goal-oriented image of the future for our society?

We know that it would. This was shown during the years of the Apollo program when we were committed to landing a man on the Moon. The result of that vision played a major role in a decade of economic health, work enough for all who wanted to work, a sense of destiny, of the acceptance of a massive challenge at which we were determined to succeed.

The acceptance of the challenge of the "high frontier" and the commitment to space industrialization, exploration, and settlement is the kind of vision that moves societies, a vision that doesn't leave an entire generation dead on the battlefields of a foreign land. The result of the acceptance of this kind of vision by a society would be a period of economic expansion that would last as long as the frontier lasted or until the Universe has been explored and charted from one end to the other.

What would it mean if our society were healthy? There would be jobs enough for those who wanted to work. In a healthy society most are able to earn a living doing what they enjoy doing. They are able to spend a significant percentage of their income on things that make life better, that help them feel better about themselves and about what they are accomplishing. There are funds enough to support programs that are designed to aid those who are ill, disabled, disadvantaged, or retired; funds for programs to help put those people with capabilities back on their feet so that they can again be contributing members. And, since many worthwhile organizations are funded privately, individuals with the funds are able to financially commit to programs that will ultimately lead to a betterment of the human condition, to the preservation of our world and its ecology.

When a society is economically unhealthy, much of this is not possible. Government funds dry up when income tax dollars drop off, and all the money from taxes goes to support the structure that has been built up over the years. In addition, individuals don't have as much disposable income, so privately funded programs find the sources of their funding reduced. As jobs become scarce, more and more people are out of work, increasing their dependence on the government or private organizations already strapped for funds.

These are some of the reasons why positive, challenging goals lead to socio-economic health; they spark the imagination, put people to work, and create wealth beyond that needed to support them. This doesn't mean that we shouldn't spend money on welfare programs; only that we need the investment return from a program like the space program to create the funds needed for programs that support people during their times of need.

The challenge of space as a goal for our nation provides a vision of a positive, challenging future that will create the wealth and knowledge to accomplish all we can dream of and more. This vision is essential to the health and future of our society.

One final thought I'd like you to think about, deals with the impact of seeing the Earth as a globe from space. We know that some people, if not all, are greatly inspired by this view and come away appreciating just how fragile our environment is. There is a good chance that as more and more of us get this global view, the impact on how we should be treating the Earth will be spread to others and, again, the concept of "Spaceship Earth" will spread and have an extremely positive impact on the issues of ecology and conservation.

I've tried to show you why, for Earth's sake, we must work for space industrialization, exploration, and settlement. I am not convinced, however, that these will be the reasons for our acceptance of this challenge.

You all will remember what was said during the early years of flight, and I quote, "If man had been meant to fly, God would have given him wings."

I believe that God did mean for humanity to fly, but he ran out of wings just before it was our turn to get a set. We were really very lucky; the penguin got the last set, and you know how far off the ground they've gotten him.

God wondered what to do and then decided to give us something better, something that would get us farther off of the ground than any other creature on the Earth. He gave us an inquiring, creative mind that would one day invent the methods needed to fly anywhere we would want to go. And then He left us a Universe to explore.

God has given us the spirit worthy of the challenges that He sets for us and we set for ourselves. He has given us the knowledge we need to find our way, so that the results of our reaching will lead to the betterment of life here and now. As a people, both individually and collectively, we can move forward into a new day that has been shaped by the combined minds of millions of individuals concentrating on this positive future. We must speak of the power to make of the Earth, the birthplace of humanity, a garden again as we accept the challenge to explore and settle the greatest frontier of all time, the infinite frontier, the Universe itself.

To end, I'd like you to take an imaginary trip with me. Think about a time when you visited an historical site, like Jerusalem or Boston, Sutter's Fort, Gettysburg, or maybe Kitty Hawk, any place of an historical beginning. Think about the people who lived then and how they felt about the events that took place there.

Now, imagine that you're in a classroom or lecture hall in a city on a planet 25 light years from the Earth. The year is 2522 (about 500 years in the future) and the instructor has just asked the class to identify the original world of the human race. A holographic map of the galaxy appears in the air at the center of the class. You raise your hand. The instructor nods at you, and, using the joystick attached to your desk computer, you move a "flying pointer" through the stars of the galaxy to identify Sol, the sun around which the original planet Earth revolves.

The instructor congratulates you, touches a button on her chair, and the hologram view zooms into this single star. You see the ten planets as they revolve in their orbits around it. Again, she looks at you and says, "Which of these planets is the Earth?"

You point to the third planet from the Sun, and again the view zooms in. Before you is a brilliant jewel of a world. A world with a magnificent single moon, almost a twin-planet system. The beauty of this world takes your breath away. You and the rest of the class have, of course, seen many pictures of the Earth, but you never fail to feel a sense of awe and deep reverence whenever you look at it.

After a moment of silence the instructor speaks to the class: "It was only a little over five and a half centuries ago that humanity first stepped off of the planet Earth and onto another world. That world was the Earth's only moon, and looking back, the step seems to be a small one, only 200,000 plus miles, just a little over what light travels in one second. But remember, until that time (for over a quarter of a million years) humanity was not able to leave that world; it did not have the knowledge or the means.

"That moment was the most important moment in our history, the time when it became possible to explore beyond the surface of that single planet, to realize our destiny of bringing life to cold, empty worlds, to live in the Universe.

"What a time to have been alive! What I wouldn't give to have been there on the Earth at that time. Just think of the adventure, the vision, the sense of destiny that must have existed. ...

The instructor goes on, but the rest of the class is lost in her basic question: What was it like to have been alive when humanity first set foot on another world? What was it like to have been alive when humanity first set out to explore and settle the Universe?

Those are the feelings and thoughts that countless generations will have as they think back to the second half of the Twentieth Century and wonder how it felt to live during that special moment in history when humanity took that first step off Mother Earth and onto another world.

As Mecca is to the Muslim and Jerusalem is to the Christian and Jew, so the Earth will be to future generations of humanity, the place from which we sprang, the holiest of holies for all time.

I believe that we will accept this challenge, that we will ultimately go into space because it is our destiny, and the pull of that destiny is too strong to be ignored.

Author's Bio

Brad Fregger has 45 years combined experience in retailing, corporate training, publishing, and software development. He has worked in large and small companies, started three of his own businesses, and worked as a senior executive in two other startups. He is currently President/CEO of Groundbreaking Press, an author-services book publishing company. Additionally, Brad is a lecturer (professor) at Texas State University-San Marcos (Business Communications) and a member of the adjunct faculty at Franklin University (Business Ethics for Leaders) in Columbus, Ohio.

Brad taught graduate-level courses at Saint Edward's University in Austin, Texas for over five years, including one year (2002) as the Executive in Residence for the Graduate School of Management. Fregger helped develop and then teach courses in the MBA Program: Introduction to eCommerce and Managerial Communications; and the Master of Science in Organizational Leadership & Ethics: Leadership & Imagination.

He also designed and taught the on-line version of Introduction to eCommerce, which received strong reviews from the Instructional Technologies department at Saint Edward's. He also taught two other courses for both the Graduate School of Management (Human Relations) and New College (Business Communications).

Brad is a practitioner/scholar, using the skills and knowledge he has learned to amass a remarkable record of accomplishment over the past 25 years, in addition to the numerous programs he developed as founder of three major corporate-training departments (Mervyns, Atari, and Activision).

He has produced more than 50 videos, 12 audio books, over 100 consumer and business enterprise software products, including the most successful computer game in the world (Shanghai) and the most played computer game in the world (computer solitaire), and published over 50 books on a wide variety of subjects.

He has completed every project begun in the past 25 years, and, even more important, during that time not a single team member left during the development of a product. This is what Brad calls "employee retention."

Brad is an expert in many critical areas of business, from customer service to the management of technology. He's an international speaker providing programs to major companies throughout the Middle East (Tunisia, Dubai, Beirut, Bahrain, Kuwait, and Qatar), Europe, and Canada in a broad spectrum of subjects including, Effective Leadership, negotiation, project management, technical risk management, creativity, and team building.

Brad Fregger

As an author, in addition to *My Thinking Cap - Solutions for Global Crisis*, he's written four other books, *Get Things Done - Ten Secrets of Creating and Leading Exceptional Teams*; *One Shovel Full - Telling Stories to Change Beliefs, Attitudes, and Perceptions*; *Lucky That Way - Stories of Seizing the Moment While Creating the Games Millions Play*; and *Why Publish - Making the Right Choices for Your Book*. Brad is currently writing his sixth book, *The Art of Leadership - Creating and Leading Effective Teams throughout the Organization*.

In addition, Brad has published articles in professional journals, including a series on book publishing for *Sharing Ideas* magazine (Los Angeles, California), an article on career change for *Career Planning and Adult Development Journal* (San Jose, California), and he wrote a column, "On a Tangent" for *Creative Pulse* magazine (Austin, Texas).

Brad's amazing ability to complete projects on time and on budget, plus his creative management style, caught the attention of Tom Peters (*In Search of Excellence*), who then featured Brad in his book, *Liberation Management*.

Brad is especially skilled in facilitation toward solving motivational and managerial communication issues. Brad believes that ineffective communication is at the heart of most personnel issues, but that these issues are often not solved by learning listening techniques.

"Courses that teach listening skills can be a waste of time. Why? Because we already know how to listen, we just choose not to. The secret is learning why we make this choice and why it's important to make other choices."

Brad holds a Master's Degree in Futuristics (San Jose State University). His speech, "Earthward Implications of Cosmic Migration," was given at the American Astronautical Society's proceedings in honor of the tenth anniversary of Apollo 11's landing on the Moon. He is a frequent guest on radio-talk shows across the nation, usually discussing the future of our society in the areas of genetic engineering, space travel, virtual reality, and extinction-level events.

Brad and his wife/business partner, Barbara Foley, live in the Texas Hill Country south of Austin.

Brad can be contacted at:
Website: www.groundbreaking.com
Email: brad@groundbreaking.com

Printed in the United States
115388LV00001B/138/P